CLYST VALE

COMMUNITY COLLEGE

Before Birth

Before Birth

Richard Dryden

*Department of Anatomy,
University of the West Indies,
Jamaica*

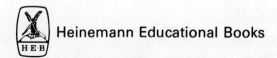
Heinemann Educational Books

Heinemann Educational Books Ltd

LONDON EDINBURGH MELBOURNE AUCKLAND TORONTO
HONG KONG SINGAPORE KUALA LUMPUR NEW DELHI
NAIROBI JOHANNESBURG LUSAKA IBADAN
KINGSTON

ISBN 0 435 60225 X
© R. J. Dryden, 1978
First published, 1978

Published by Heinemann Educational Books Ltd
48 Charles Street, London W1X 8AH

Phototypesetting by George Over Ltd., London and Rugby
Printed in Great Britain by Butler & Tanner Ltd., Frome and London

Preface

Embryology is a fascinating subject, yet to sixth-formers and medical students the word alone is sufficiently unpopular to elicit a shudder. I feel that this paradox is largely due to the way embryology is presented by teachers and the traditional texts. We seem to have inherited a description of development based on wax models and old-fashioned germanic-sounding labels, with an emphasis on discontinuous stages rather than the overall continuous flow of change.

My own awakening to the significance of embryology came as I waited for an interview at a medical school. To pass the hour or so in hand I browsed among the pickled diseases of the pathology museum and came eventually to a section devoted to congenital malformations. Scanning along the row of jars containing twisted limbs and distorted faces my initial feeling was one of revulsion tinged with dismay — dismay that anything as new as a newborn baby should be other than perfect.

Since then my interest in development has survived and grown. My attitude has changed, however, and the original feeling of loathing for abnormal development has been superseded by a more positive admiration of the process of normal development, which, after all, occurs in the majority of instances.

The first section of this book gives a brief introduction to the subject of embryology. The second section reviews prenatal and neonatal development as it is understood today. Obviously, with the limited space available, it was necessary to make a subjective selection of material for inclusion. It will be noticed that the order of presentation is a little different from that in other texts on embryology. For example, gametogenesis (formation of eggs and spermatozoa) has been relegated to the end of the section, instead of providing a starting point for the description. This is to reduce the risk of boredom setting in before the word 'embryo' is even mentioned. To me, conception seems to be a more natural place to begin the story.

The third section of the book is concerned with how and why development sometimes goes astray, and a relatively detailed account is given of spina bifida to illustrate the way in which different approaches may be integrated. The last section dips into the more abstract, theoretical aspects of development and the device of model building.

Thus, after a conventional introduction to the subject, the reader is pushed rapidly towards the deep end — hopefully not before he knows enough for survival, but perhaps quickly enough to make him throw up his arms and complain. The aim is to draw attention to the diverse lines of study that may one day contribute to our understanding of prenatal development.

My especial thanks are offered to Mr B. King for his ideas which helped so much during my excursion into the realm of model making and to Graham Taylor of Heinemann Educational Books for his help and encouragement in the later stages of preparing the book.

1977   R. J. D.

Contents

		Page
Preface		v
1.	An Introduction to Embryology	1
	Historical perspective	1
	Periods of development	5
	Summary of prenatal development	8
	Embryonic cells and human societies	16
2.	A Description of Normal Development	18
	Formation and implantation of the blastocyst	18
	Twinning	24
	The embryonic period	27
	The nervous system	35
	The heart and major blood vessels	44
	The face	53
	The eye	56
	The ear	59
	The digestive system	61
	The respiratory system	70
	The musculoskeletal system	72
	The urinary system	78
	The genital system	81
	The foetal period	88
	The amnion, chorion, and placenta	92
	Birth and the neonatal period	99
	Gametogenesis	103
3.	Abnormal Development	111
	The incidence, causes, and study of abnormal development	111
	The problem of spina bifida	121
4.	Theoretical Aspects	131
	Understanding development	131
	Model building	141
	Developing a model of development	146
Glossary		161
Bibliography		168
Index		173

List of Figures

1. An Introduction to Embryology
Figure 1   Drawing of a preformed individual residing in the head of a spermatozoon.
Figure 2   The first experiments in embryology.
Figure 3   An individual's life span subdivided into periods of development.
Figure 4   The dedifferentiation of leukemic cells.
Figure 5   The 1st week of development — from fertilization to implantation.
Figure 6   A fully implanted blastocyst about 12 days after conception.
Figure 7   Ectodermal aspects of human embryos at 18 days and 20 days.
Figure 8   Embryo at the end of the 3rd week.
Figure 9   An embryo towards the end of the 4th week.
Figure 10   Embryos at 4—5 weeks.
Figure 11   An embryo in the middle of the 6th week.
Figure 12   An embryo at the end of the 7th week.
Figure 13   A 6-month foetus with its membranes.

2. A Description of Normal Development
Figure 14   Entry of the spermatozoon into the ovum.
Figure 15   Hamster spermatozoa.
Figure 16   Human mitotic chromosome.
Figure 17   Effect of partial subdivision of the inner cell mass.
Figure 18   The primitive streak.
Figure 19   The three embryonic germ layers.
Figure 20   The positions of the buccopharyngeal membrane and the cloacal membrane.
Figure 21   Ultrastructural appearance of the notochord in transverse section.
Figure 22   Formation of the neural plate in relation to the notochord.
Figure 23   Subdivisions of the embryonic mesoderm.
Figure 24   The mechanism of neurulation.
Figure 25   Cells of the neural crest.
Figure 26   The developing brain region of a hamster embryo.
Figure 27   Cell division in the neural tube.
Figure 28   The developing spinal cord.
Figure 29   Development of the brain.
Figure 30   The ventricular system of the brain and the flow of cerebrospinal fluid.

Figure 31 Early stages in the development of the heart.
Figure 32 The pattern of blood circulation in a 4-week embryo.
Figure 33 Subdivisions of the primitive heart tube, and formation of the cardiac loop.
Figure 34 Septation of the atrium and ventricle.
Figure 35 Formation of the spiral septum in the bulbus cordis.
Figure 36 Development of the major arteries.
Figure 37 The developing venous system.
Figure 38 Circulation of blood in the foetus.
Figure 39 Development of the face.
Figure 40 The face and hands at 12 weeks.
Figure 41 The face and hands at 20 weeks.
Figure 42 Development of the eye.
Figure 43 Cyclopia.
Figure 44 Development of the ear.
Figure 45 Initial stages in the development of the digestive system.
Figure 46 The digestive system in an embryo at the end of the 4th week.
Figure 47 The embryonic pharynx.
Figure 48 Rotation of the midgut.
Figure 49 Development of the respiratory system.
Figure 50 Subdivisions of the somite.
Figure 51 The pattern of ossification in a 4-month foetus.
Figure 52 Development of intersegmental vertebrae from sclerotomal cells.
Figure 53 The developing urinary system.
Figure 54 The urogenital system in a 13-week human foetus.
Figure 55 Horseshoe kidney.
Figure 56 Formation of the primitive gonad.
Figure 57 Development of the male reproductive system.
Figure 58 Development of the female reproductive system.
Figure 59 Development of the external genitalia.
Figure 60 A 10-week foetus and placenta.
Figure 61 A 12-week foetus.
Figure 62 A 20-week foetus.
Figure 63 General organization of the placenta.
Figure 64 Different placental forms.
Figure 65 Delivery of the baby.
Figure 66 The pattern of blood circulation in the adult.
Figure 67 Transformation of a spermatid into a spermatozoon.
Figure 68 Stages in oogenesis.
Figure 69 Chromosomes from a human cell undergoing mitosis.

3. Abnormal Development
Figure 70 A normal distribution.
Figure 71 The external appearance and types of spina bifida.
Figure 72 Control of hydrocephalus.
Figure 73 Hypotheses relating to the embryology of spina bifida.
Figure 74 A comparison of embryos from different species.
Figure 75 Early secondary changes in spina bifida.

Figure 76   A bacterium that has been ruptured by osmotic shock to reveal its long chromosome.

4.   Theoretical Aspects
Figure 77   Two interpretations of chromosomal structure.
Figure 78   The Watson and Crick model of DNA.
Figure 79   Replication of DNA.
Figure 80   Punctuation of the genetic code.
Figure 81   Transcription of genetic information.
Figure 82   Distribution of ribosomes.
Figure 83   Protein synthesis.
Figure 84   Regulation of protein synthesis.
Figure 85   Subdivision of the cytoplasm of the zygote during cleavage divisions.
Figure 86   Mitochondrial form and structure.
Figure 87   A comparison between the potential of a zygote and the mean potential of the tissues of an adult.
Figure 88   The developmental hill.
Figure 89   The effect of fertilization on developmental potential.
Figure 90   Changes in the developmental potential of gametes.
Figure 91   Developmental curves.
Figure 92   A continuation of the hill analogy.
Figure 93   The waterbed analogy.
Figure 94   The developmental field after fertilization.
Figure 95   The developmental field at a later stage.
Figure 96   A ripple passing through the developmental field.
Figure 97   Genetic intervention.

Glossary
Figure 98   Terms used when describing positional relationships in embryos and adults.

Idle Beauty

'But I maintain that nature is our enemy. We have to be continually resisting her efforts to reduce us to the level of brutes. Whatever on this earth is seemly, comely, delicate, and poetic, we owe not to nature but to man, to the human brain. Thanks to us, thanks to the poets, who have sung and interpreted and praised it; to the artists, who have idealized it, to the scientists, who, in self-delusion, have explained it and set forth ingenious reasons for its various phenomena, creation is redeemed by some touch of grace and beauty, some hint of indefinable charm and mystery. Nature has created none but rudimentary beings, swarming with germs of disease, doomed, after a few short years of animal development, to an old age hideous with all the infirmities and disabilities of human decrepitude. Mankind, it seems, is created only to reproduce itself in squalor, and then to die . . . It is as if a cynical and perfidious creator had schemed to prevent man from ever ennobling, exalting, and idealizing his relations to woman. But man has invented love, not such a bad rejoinder to the wiles of destiny, and has so adorned it with poetic fancies, that woman often forgets the gross facts. Some of us, who cannot accept these delusions, have invented vice and brought it to a fine art, which is another way of tricking providence and rendering homage, however tainted, to beauty. But ordinary mortals beget children, like a pair of animals, mated by law.'

Guy de Maupassant

1 An Introduction to Embryology

Historical Perspective

Embryology is a four-dimensional subject. Those studying it are required to juggle not only with the three dimensions of shape and form but also with the factor of time. Understanding a complex structure is difficult even when it is static, but the difficulties in studying transitory stages in an essentially unbroken sequence are obviously greater. However, the rewards are correspondingly great, and it is often only as a result of developmental studies that the final arrangement can be comprehended. But awareness of the flow of structural changes is only the first step, and the questions that currently dominate embryological discussions are concerned with how these changes are initiated and controlled: embryology is now increasingly involved in an analysis of organization.

Ideally, an introduction to a subject will provide a firm basis from which to start, a framework to which can be fitted the numerous facts and concepts as they arise. In the case of embryology the best introduction is provided by the story of its own development as a field of study. The path taken historically was deflected and rutted by philosophical prejudices, and to some extent these still exist today. For example, it is still believed by some that the perceptual experiences of the mother are passed on to the foetus in her womb and that human embryos relive phylogenetic evolution by becoming successively a fish, an amphibian, and then finally a human being. First, then, it will be helpful to see how some of these points have been resolved.

As in other fields of study, it was the Greeks who first collected and classified information on development. The outcome was a mixture of facts and fallacies. It was appreciated that the foetus developed from matter contributed by both mother and father, and that the heart developed before the fingernails, but it was also stressed that a mother wanting a beautiful child should study statues and pictures of acknowledged beauty. This theory of 'maternal impressions' seems to us unlikely, and implies that the mother passes on acquired characteristics to her child. However, it cannot be completely dismissed in view of the recent observation that women who suffer prolonged stress during pregnancy tend to have smaller, more irritable babies than usual.

Aristotle's pronouncements in particular carried a great deal of authority and influence, although by present day standards he underestimated the importance of the female's contribution. He proposed that semen provided the organizational force behind development while the mother provided only the raw materials for this process of creation. However, one passage in Aristotle's book *Generation of Animals* puts forward a view of development that is still valid: 'In the early stages, the parts are all traced out in outline, later on they get their various colours and softnesses and hardnesses, for all the world as if a painter were at work on them, the painter being Nature.' This corresponds accurately with the picture given by microscopy: the various organs are first fashioned in a simplified way before becoming structurally and functionally complex.

It was not until Renaissance times that Aristotle's theories were elaborated further. Fabricus, a student of Fallopius in the sixteenth century, was particularly interested in direct observations on embryonic and foetal structure and published works describing the development of birds and mammals, including man.

In the seventeenth century William Harvey, whose demonstration of the circulation of blood laid the foundations of experimental biology, made an important contribution to embryology. He demonstrated that the pale disc in the yolk of the hen's egg gives rise to the embryo and dismissed the Aristotelian idea that female contributes substance and male provides form, suggesting instead that both components should be given equal credit. However, his views on conception were somewhat confused because he was unable to detect any mixing of gametes at coition.

His sound observations were liberally sprinkled with metaphysical considerations. For example, he looked upon the egg as the point from which all life originates and at the same time thought that the purpose of all life is to produce more eggs.

Although he proposed that the organs arise successively during development, a similar outlook to that of Aristotle and a process currently known as *epigenesis*, Harvey considered that in some animals there is a degree of *preformation* of structure within the egg, in which case there is merely growth in size during development (*see Figure 1*). This idea was taken up in 1673 by Malpighi who, using the then recently developed microscope, believed that he could see pre-existing parts of the adult in the unincubated hen's egg. Apparently, the eggs he studied had previously been lying in the sun in Bologna, so it is possible that some embryonic development had occurred before he opened them. As a result of his work, the preformation theory was taken to absurd lengths. Among other things, there were speculations about the necessary size of the first rabbit within whose ovaries all the rabbits from beginning to end of time were represented. There was similar speculation about Eve.

In the eighteenth century the preformation hypothesis was questioned by Caspar Wolff. He observed that development occurs by gradual differentiation of originally homogeneous material and that organs such as the intestines and central nervous system are formed by the folding of tissue layers into tubes.

In modified form, the controversy between those who propose preformation and those who support epigenesis still continues today. Although it is clear that during embryonic development new structures and organs appear by epigenesis, it can be argued that all the information required to produce these changes is stored in the chromosomes within the cell nuclei. Thus, from the standpoint of the geneticist development is a consequence of preformation.

Following the advance made by Wolff, there was a lull in which philosophical discussion took the place of scientific study. Goethe suggested that natural

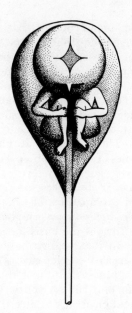

Figure 1. Drawing of a preformed individual residing in the head of a spermatozoon. [Copied from a drawing made by Hartsoeker early in the eighteenth century]

phenomena represent modifications of an idea in the mind of the Creator. Kant did not reject natural science so completely, but unfortunately the prevailing metaphysical outlook discouraged experimental verification of ideas. Embryology returned more positively to the realms of science some fifty years later when Pander, working in Wurtzburg, first demonstrated the existence of three primary tissue layers — *germ layers* — in chick embryos. This significant finding was explored more deeply by von Baer, who studied embryos from various species. He was able to show similar processes in all the forms he studied and went on to emphasize the importance and value of embryology in studies of comparative anatomy.

Although von Baer warned against making too many conclusions from the observation that species were comparable at some stages of development, it became fashionable to stress the similarities between the embryos of higher forms and the adults of lower forms. Rathke's findings that mammalian embryos possess transient gill arches during development, and that both vertebrates and invertebrates develop from three germ layers, reinforced this misguided approach. The idea culminated in Haeckel's biogenetic law, which states that *ontogeny recapitulates phylogeny*, or, to put it another way, he believed that the development of an individual demonstrated the evolutionary history of the species to which it belonged. To some extent this is true. The early development of all vertebrate embryos is strikingly similar, and it is only later that the developmental pathways of the different species diverge. The stage at which the precursor of the central nervous system is formed is particularly comparable between species. However, a fundamental error was made in considering that the embryo passed through stages that were equivalent to adult forms of animals more primitive in an evolutionary sense. The observation that the embryos of man and fish closely resemble each other implies that common developmental mechanisms have been retained by both species, presumably because they have proved to be the most

4 Before Birth

successful, and does not indicate that a human embryo becomes a fish for a short time during development. The interspecies similarity of early development allows us to extrapolate with caution from embryos of laboratory animals to human embryos, which are obviously not so amenable to study. However, there are limits to these similarities, as for example are demonstrated by the differing responses of experimental embryos to drugs harmful in man.

Let us return to Haeckel. In 1872 he pointed out the similarity of all gastrulae and even invented, with illustrations, a free-swimming 'gastrea' that was a common ancestor to all animal forms. The biogenetic law immediately became popular and soon permeated into studies of cultural and psychological development. Unfortunately, in embryology it tended to restrict studies purely to evolutionary problems. Towards the end of the nineteenth century, His, Goette, and Rauber put forward mechanical explanations of development, but they attracted little attention because of the predominant emphasis on phylogeny.

Experimental embryology, which provides the basis for current knowledge, began when Roux set out to determine whether the distribution of the contents of a fertilized egg dictates subsequent patterns of development. With a hot needle he killed one of the cells *(blastomeres)* in amphibian embryos at the two-cell stage and found that the viable cell developed into only half an embryo *(see Figure 2*a). This result seemed to support the preformation theory. Later, Driesch and others came to a different conclusion. In their apparently similar experiments, in which the blastomeres were completely separated by pulling them apart with a loop of hair, each cell developed into a complete and normal embryo *(see Figure 2*b). This suggests that the developmental fate of particular cells is not fixed (determined) at the beginning of development, and it seems that initially cells possess a greater potential than they will generally require. Thus the differentiation of a cell depends on its position in relation to the whole embryo. In Roux's earlier experiments, the presence of the coagulated blastomere must have prevented the viable cell from realizing its full potential.

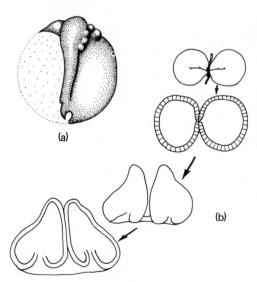

Figure 2. The first experiments in embryology. (a) Roux obtained a half-embryo after killing one blastomere with a hot needle. (b) Driesch separated the first two blastomeres of a sea urchin embryo by tying a loop of hair between them and obtained two complete embryos. [After Roux (1888), *Virchows Arch.*, *114*, 419 and Driesch (1891), *Z. wiss Zool.*, *53*, 160]

The factors that regulate normal embryonic development have been investigated by increasingly precise techniques such as tissue grafting and extirpation, and a number of principles have emerged. Spemann, Mangold, Harrison, and others developed the concept of induction, showing that one region of an embryo may dictate or modify the development of an adjacent region, probably by way of a chemical mediator passing from one tissue to the other. Embryologists today envisage that a chain reaction of inductive processes contributes to the orderly sequence of changes that occur during early development.

From this brief review it can be seen that the study of embryology reflects to some degree the nature of the processes being analysed. Our knowledge of developmental phenomena has only recently begun the transition from simple description of structure to clearly defined studies of developmental mechanisms. Thus, returning to Aristotle's analogy, we have sketched in the outlines of development and are now beginning to insert a few details. However, prenatal development, which previously must have seemed as enigmatic as a beautiful smile in the dark, has lost none of its beauty and only a little of its mystery in the light of scientific study.

Periods of Development

From conception to death the structure and activity of the human individual changes continuously, but when considering these changes it is helpful to subdivide the sequence into discrete stages. For example, birth is a clearly defined transition from a totally dependent, almost parasitic mode of life to an increasingly independent existence. Thus to some extent we can consider *prenatal* and *postnatal* development in isolation, leaving until later the more intricate task of correlating the findings. Smaller subdivisions of the sequence are often harder to define, since transitional periods are usually fleeting events in the flow of development and may involve only some regions of the organism at a given time. However, the following somewhat arbitrarily chosen steps are widely used for descriptive purposes (*see Figure 3*).

The *pre-implantation period* begins with fertilization of the ovum by a spermatozoon. The zygote divides repeatedly to form a growing community of cells, and at the end of the 1st week the hollow *blastocyst* buries itself into the wall of the mother's uterus. This event is known as *implantation*.

In the subsequent *embryonic period*, which extends to the end of the 2nd month, the basic organs and systems of the body are established. Only 8 weeks after conception the embryo already has a remarkably human external appearance, possessing eyes, ears, a nose, a mouth, and well-developed hands and feet. Internally there is a blood circulatory system, a nervous system, a digestive tract, and primitive urinary and genital systems. However, the embryo is diminutive in size (3 cm long), and of all its systems only the heart and circulatory system can be considered functional.

This functional immaturity of the majority of the newly-formed structures precludes survival outside the mother's uterus at this stage, and a comparatively long period of maturation and growth must occur before birth. This period is

6 Before Birth

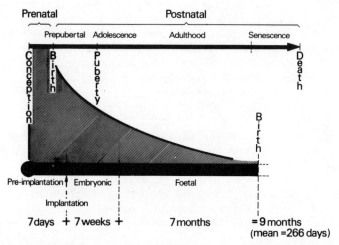

Figure 3. An individual's life span subdivided into periods of development.

known as the *foetal period* and normally lasts some seven months. The chance of postnatal survival decreases if birth occurs markedly before or after this time, although some babies born up to two months prematurely can survive.

When the intimate physical relationship between the mother and the baby is broken at birth, the baby is still very dependent and requires care and protection. A long period of further development must elapse before complete independence is gained, and during this time it is clear that environmental factors can greatly alter the mental and physical characteristics of the individual.

The period between birth and puberty is referred to as *prepubertal* development. At puberty the individual reaches reproductive maturity, and this is followed by the periods of *adolescence* and *adulthood*. Towards the end of adulthood there is a period of *senescence*, when a gradual decline of mental and physical powers occurs before death. The mechanism of ageing is still unknown, but it is interesting to note that in cancer, which in one form or another often contributes to death in old age, some cells seem to revert to an embryonic type of structure and begin to divide rapidly, shrugging off the normal controlling influences exerted by the body. This dedifferentiation of cancer cells is thought by some to represent a reversal of the process of embryonic differentiation. (*See Figure 4.*)

The following sections are concerned specifically with prenatal development, but on occasions it will be necessary to continue the description into the early postnatal period. For example, when development of the circulatory system is being considered, the description would be inadequate if it did not include a mention of the dramatic changes in blood flow that occur during and after birth. Because of the complexity of prenatal development and the tendency for several events to occur simultaneously, especially during the embryonic period, it will be necessary to fragment the account into a number of subsections, with the result that some of the essential interactions that occur during development may be obscured. As a partial insurance against this a brief summary of the whole sequence will now be given.

An Introduction to Embryology 7

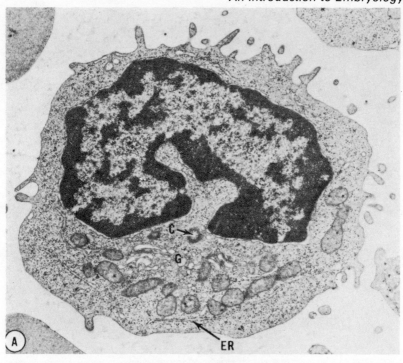

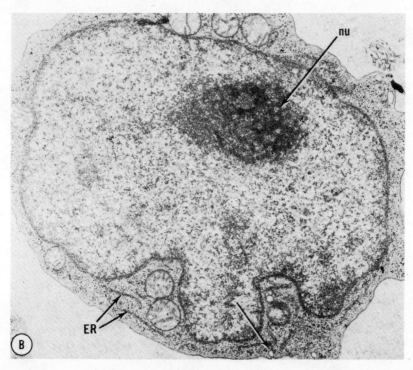

8 Before Birth

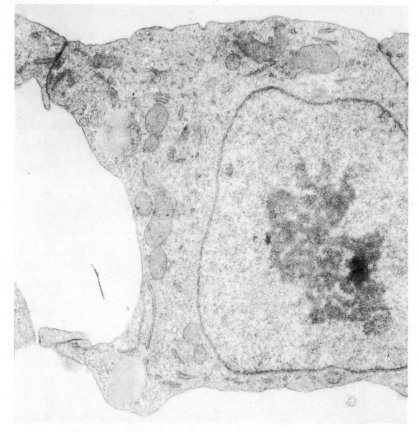

Figure 4. The dedifferentiation of leukemic cells. (A) The ultrastructural appearance of a normal lymphocyte; (B) a lymphocyte from a patient with leukemia; and (C) an endodermal cell from an early chick embryo. Note the similarity of the leukemic cell to the undifferentiated embryonic cell. The nucleus of the normal lymphocyte displays clumps of condensed heterochromatin adjacent to the nuclear envelope, a feature that is characteristic of highly differentiated cells in which there is little protein synthesis. [(A) and (B) from Anderson, D. R. (1966) *Journal of Ultrastructural Research,* **2,** Supplement 1]

Summary of Prenatal Development

Pre-implantation period

Development of a new individual begins when the two specially prepared gametes contributed by the parents meet and fuse. This phenomenon of fertilization occurs in the mother's reproductive tract close to the ovary. The resultant cell embarks on a sequence of changes that is characterized by a remarkable increase in structural complexity and functional capability, changes

An Introduction to Embryology

that are referred to as differentiation. There is also dramatic growth. As the initial cell divisions occur a ball of cells is produced, and this is carried passively down the uterine tube until it reaches the lumen of the uterus. (*See Figure 5*.) On entering the uterus some 5 – 8 days after fertilization, the ball of cells becomes reorganized to form a hollow sphere. Internally, a small group of cells segregates from the enveloping sphere and begins to develop into the embryonic body and associated membranes.

Implantation

At this stage of development the conceptus makes contact with the internal lining of the uterus and becomes attached to it. The process of implantation then begins. The conceptus invades and steadily buries itself in the maternal tissue, which in preparation for this event is glandular, oedematous, and richly supplied with blood. Implantation is completed shortly before the end of the 2nd week. The outer cellular sphere becomes modified to produce a vital supporting tissue – the chorion – which establishes an intimate relationship with the maternal tissues. In one region the tissues of the mother and conceptus begin to co-operate in the formation of the placenta, an efficient organ of exchange between the mother and the conceptus.

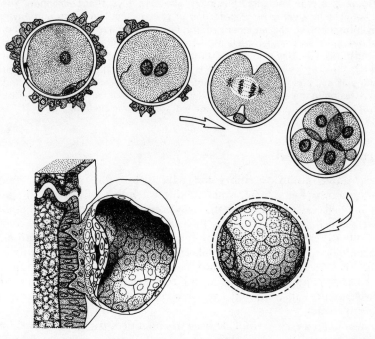

Figure 5. The 1st week of development – from fertilization to implantation. Cleavage divisions give rise first to a ball of cells still contained within the zona pellucida *(upper left)*. The ball then becomes transformed into a hollow sphere *(lower right)* that attaches to the lining of the uterus and begins to implant *(lower left)*. By this time a small group of cells called the inner cell mass can be distinguished within the blastocyst.

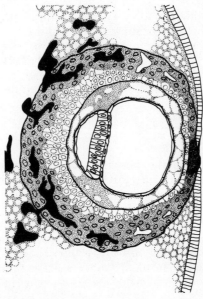

Figure 6. A fully implanted blastocyst about 12 days after conception. The trophoblastic shell is now composed of two distinct layers: an outer syncytiotrophoblast, and an inner cytiotrophoblast. Lacunae of maternal blood, shown here as irregular areas of black, are becoming intimately associated with the syncytial layer. Internally the inner cell mass has differentiated to form a bilaminar embryonic disc, which is continuous laterally with the linings of the yolk sac and the amniotic cavity. Fluid-filled spaces are appearing and coalescing in the interval between derivatives of the inner cell mass and the trophoblast, thereby establishing the extraembryonic coelom.

Embryonic period

The segregated group of embryonic cells within the implanted conceptus becomes organized as a membrane crossing the lumen of the hollow sphere of supporting cells, and before long the membrane can be seen to be composed of two layers of cells (*see Figure 6*). The thicker of the two layers is called ectoderm, and the other is called endoderm. The amniotic cavity develops in relation to the ectodermal layer, and this steadily enlarges to become a large, fluid-filled cavity allowing freedom of growth and movement to the embryo and foetus. The cavity of the conceptus lying in relation to the endoderm becomes surrounded by a bag-like structure called the yolk sac, which later contributes to the formation of the digestive tract.

The two-layered embryo quickly becomes modified to a three-layered form by a process of migration of some of the ectodermal cells into the region between the ectoderm and the endoderm. This intermediate third layer is called mesoderm. From the beginning of the 4th week of development to the end of the 8th week, the three layers undergo complex changes, which result in the formation of all the major organs and systems of the body. (*See Figures 7–12*). However, although structural organization is established, very few of the organs are functional at this

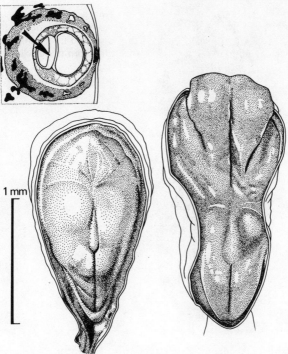

Figure 7. Ectodermal aspects of human embryos at 18 days *(left)* and 20 days *(right)*. The primitive streak appears first, followed by the part of the neural plate that will form the brain region. *Upper left:* Diagram indicating the part of the conceptus shown in the other drawings. [Figures 7–13 after Blechschmidt (1961), except 10b]

stage. The main exception to this general lack of functional maturity is the cardiovascular system comprising the heart and blood vessels. These play an essential role in transporting materials both within the embryo and to and from the placenta. The placenta enables the embryo to obtain oxygen and nutrients from the mother and at the same time to eliminate waste products into the maternal bloodstream, whilst maintaining a physical separation between the two bloodstreams. The structure that initially links the embryo with the surrounding trophoblast — the body stalk — becomes modified to form the long and tortuous umbilical cord. This contains the umbilical vessels carrying blood between the embryo and the placenta.

The embryonic period of organ building is typified by the folding of cell layers and the migration of individual cells. These activities are set against a background of continuous cell division and sporadic cell death. The central strip of the ectodermal layer folds to form a tubular central nervous system, and the remainder becomes the outer covering of the embryo. The endodermal layer folds to form the digestive tract, from which a number of accessory organs and the respiratory system branch out. The cells of the more loosely organized mesodermal layer form the heart and blood vessels, contribute to structures developing from the other germ layers, and form aggregations between the outer ectodermal

12 Before Birth

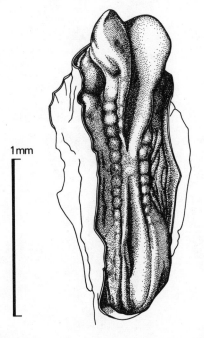

Figure 8. Embryo at the end of the 3rd week. Fusion of the neural folds has occurred in the middle of the embryo and from there will progress cranially and caudally. A row of segmental somites is apparent on each side of the neural plate.

covering and the endodermal digestive tract. Some of these aggregations are segmentally arranged and called somites. The somites form muscles, skeletal structures, and the deep dermal layer of the skin. Other aggregations of mesodermal cells develop into the urinary and reproductive systems. These latter systems are at first closely interlinked, but some degree of separation soon appears. Separation eventually becomes total in the female, but it is less complete in the male. Sex differences are not immediately apparent in the embryo, and the gonads are seen to pass through an indifferent period during which development is the same in both male and female. After the 6th week, however, differences begin to emerge.

The various systems and structures develop at different rates, and their times of first appearance in the embryo also differ, so the embryonic period may be considered as a collection of overlapping and yet closely integrated sequences of development. Harmful environmental factors, known as teratogens, are particularly potent during the embryonic period, and as might be expected there is a significant relationship between the stage of embryonic development at the time of exposure and the types of abnormality that result. Each organ seems to pass through a 'critical period' when it is highly susceptible to damage by teratogens, and this knowledge can be used to interpret not only abnormalities but also some of the processes of normal development.

By the end of the embryonic period the embryo has a recognizably human form externally (see Figure 12). Internally all the major systems are present. Much of the skeletal system is represented by cartilagenous models of the future bones, and the process of ossification begins, marking the transition into the foetal period. Ossification continues until early adulthood, when growth of the skeleton ceases. The muscles show the first signs of activity at the end of the embryonic

An Introduction to Embryology 13

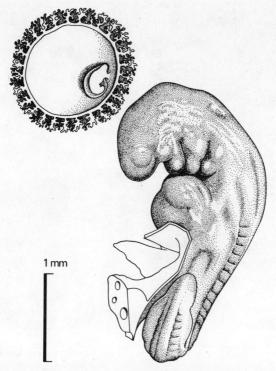

Figure 9. An embryo towards the end of the 4th week. The neural tube is complete, the developing eye is visible, pharyngeal arches are prominent in the neck region, and the large bulbous heart region is placed within the curve made by the embryonic body. *Upper left:* Diagram illustrating the general proportions of the whole conceptus at this stage.

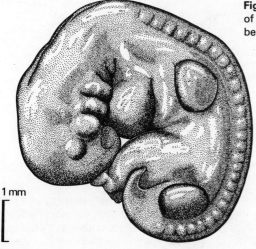

Figure 10. (a) Embryo at the beginning of the 5th week. The limb buds are beginning to develop.

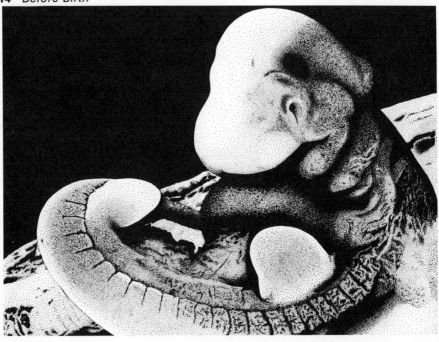

Figure 10. (b) A 5-week human embryo as it appears in the scanning electron microscope. [after Lennart Nilsson, Karolinska Institute, Stockholm]

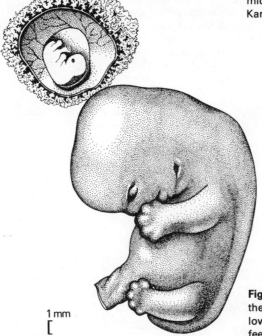

1 mm

Figure 11. An embryo in the middle of the 6th week. The external ear is visible low in the neck region, and hands and feet are paddle-shaped at this stage.

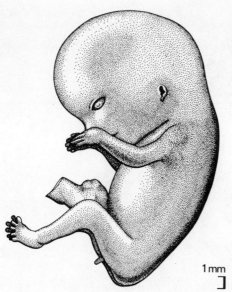

Figure 12. An embryo at the end of the 7th week. The external form is clearly human.

1 mm

period and become increasingly dependent on their relationship with the central nervous system via the peripheral nerves. Muscles deprived of a normal innervation develop abnormally. Most of the muscles of the body, with the exception of those found in visceral structures or blood vessels, originate from parts of the segmental somites and the walls of the pharynx. Each somite is linked with the central nervous system by a peripheral nerve, and when the components of the somite migrate to their definitive positions the relationship with the original nerve is retained. Thus a segmental pattern of innervation can be detected even in an adult, in both the motor supply to muscles and the sensory innervation of the dermis. The embryonic brain already shows many recognizable features but at this stage is almost totally nonfunctional. Maturation of the brain is not complete even at birth, although by that time the essential regulatory mechanisms have been established. The special sense organs (the eye and the ear) are also structurally quite advanced in the embryo, but once again a long period must elapse before they become fully functional and before the regions of the brain that interpret the information they provide become operational. The internal reproductive system is clearly delineated, and the external genitalia are obviously male or female, but functional maturity is not, of course, reached until puberty.

Foetal period

From the 9th week until birth — the foetal period — the various systems of the body differentiate and mature sufficiently to enable survival outside the uterus. It is a period marked also by phenomenal growth. The length from the crown of the head to the buttocks increases from about 3 cm to 35 cm, and the weight increases from 4 g to over 3 000 g. The proportions of the body change towards those of the newborn, the head becoming proportionately smaller and the limbs longer. (*See Figure 13.*) A number of reflex activities are rehearsed before birth in preparation

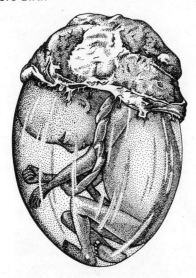

Figure 13. A 6-month foetus with its membranes.

for the needs that arise after birth. For example, the important swallowing reflex is perfected *in utero,* and the groups of muscles that produce the movements of respiration have also been co-ordinated by the time demands are made of them.

Birth

The events that precipitate birth are not clearly understood but probably include hormonal changes and a loss of placental efficiency. The mother's uterine musculature contracts very forcibly and with increasing frequency until the foetal membranes — the amnion and chorion — rupture and the baby is expelled via the birth canal. The umbilical cord is tied and cut, and the membranes, including the discoid placenta, are expelled from the uterus soon afterwards. Thus the baby begins its struggle towards an independent existence.

Embryonic Cells and Human Societies

Living in a technological society we have become used to the idea of taking raw materials and fashioning them into complex forms. However, the organs and systems of the body clearly arise in a different way. The raw materials of life are fashioned and controlled from within the organism and not from outside. Thus structure and organization *evolve* from an apparently simple starting point — the zygote. There is in addition the necessity for the organism to be viable at every stage in this sequence.

It is now known that within the zygote there is a large quantity of information. It is stored mainly in the chromosomes and copied each time cell division occurs, identical copies being passed on to the resulting daughter cells. This information probably includes a set of instructions to guide development, but there are reasons for thinking that development is not entirely due to genetic dictates. The embryo is very sensitive to the environment in which it is growing, and it is more realistic to consider that development involves an interaction between genetic and environmental factors. This point will be discussed in the last section of the book. However, it is important that the description of development in the following section should be read with some idea of this co-operation in mind.

Fortunately there is a useful analogy available that can illustrate this concept in terms of everyday experience. It is based on the observation that there are a number of similarities between the development of organization in a community of embryonic cells and the social development of human communities. If conditions are favourable, the zygote and suceeding cells divide repeatedly until the appropriate multicellular form is achieved. In the adult body most of the component cells show a striking degree of specialization, in both structure and function. However, in achieving this degree of specialization individual cells lose their ability to change into a cell with a different structure, and they become dependent on other cells for their maintenance. They often become unable to divide. A similar type of specialization, or division of labour, can be seen to arise in human societies. As a consequence the group as a whole becomes more versatile and adaptable, but each member of the society becomes increasingly dependent on the others for survival. Thus in both embryological development and social development, increasing complexity and flexibility are correlated with increasing interdependence within the group, whether it is cells or people. But — and this is the important point — although the viability of both types of community depends largely on the effectiveness of its internal activities, it is impossible to separate these activities from the environment in which the community exists, and the internal organization must have sufficient adaptability to meet changing internal and environmental demands. It could of course be argued that embryos have been rather more successful at this than human societies have.

2 A Description of Normal Development

Formation and Implantation of the Blastocyst

Fertilization

Fertilization is the fusion of two specialized cells to form a single cell containing enough potential to build a new individual. The cells that combine — the *gametes* — are very different from each other morphologically. The small motile spermatozoon, the male gamete, is essentially a parcel of genetic information to which is attached a short-lived power plant. The ovum is very much larger, and in addition to its genetic store it possesses a substantial quantity of cytoplasm. It is coated externally by a tough elastic capsule called the *zona pellucida*, to which there may still be a few ovarian cells adhering. Although different morphologically, the two gametes show an important similarity: the nucleus of each contains only half the usual number of chromosomes. Thus there has been a *reduction* of genetic material during formation of the gametes. The need for this reduction is clear: repeated fusion of cells in which no reduction has occurred would lead after only a few generations to an accumulation of genetic material so great that normal functioning of the cells could no longer continue. In the human, most cells of the body contain forty-six chromosomes. Two of these chromosomes play an important role in determining the development of sex-specific structures and behaviour in the body and are called *sex chromosomes*, while the other forty-four control the development and function of systems in a more general way and are called *autosomes*. In the male the sex chromosomes are different in appearance and are designated X and Y; in the female they are identical and are both of the X type. During formation of the gametes, the sex chromosomes become separated and each gamete receives only one: either an X or a Y chromosome if it is a spermatozoon, or a single X if it is an ovum. Thus fusion of the gametes at fertilization brings together two sex chromosomes and establishes the *genetic sex* of the individual. If the spermatozoon carries an X chromosome to the ovum, a girl is conceived; if it contributes a Y chromosome, a boy develops. However, the final morphology of the individual is dependent on a number of factors other than genetic sex. For example, hormones influence development of the body and sex-specific behaviour during certain critical periods of development. If their balance is disturbed, it is possible for the genetic sex to be masked by atypical physical development.

The physical mechanism of fertilization has been studied in detail by various techniques and in many species. The electron microscope has been particularly helpful in demonstrating the structural changes that occur during the penetration of the spermatozoon into the ovum, while delicate biochemical tests have provided clues to the enzymatic events that are involved in the spermatozoon's passage through the zona pellucida. In some species of animals it is thought that spermatozoa are lured towards the ovum by chemical attraction (chemotaxis), the chemical being secreted by the ovum. In the human context, however, chemotaxis does not seem to occur, and fertilization is left to the chance 'bumping together' of gametes. Spermatozoa have been seen to dance erratically as they drift along the uterine tube towards the ovum, and the majority pass blindly on. The small number of spermatozoa that contact the covering of the ovum initiate the next step in fertilization. First, the zona pellucida must be penetrated. It seems that on its own a single spermatozoon cannot do this and that only the joint action of several spermatozoa can overcome this barrier. They release proteolytic enzymes from a membrane-bound bag within the head of each spermatozoon, and these act on the zona pellucida, allowing one or perhaps several to reach the cell membrane of the ovum itself. However, only one penetrates the ovum within. This spermatozoon completes fertilization and provides the incentive for embryonic development to begin by completing the store of genetic information. As soon as penetration has occurred a change takes place in the cell membrane of the egg, and invasion by additional spermatozoa is prevented.

The entry of the spermatozoon into the ovum is effected by fusion of their outer membranes (see Figures *14* and *15*). Once within the ovum, the head of the spermatozoon becomes detached from the middle piece and tail and moves towards the centre of the cell, enlarging as it does so. Here it meets the nucleus of the ovum, which has also undergone preparatory changes, and fertilization is complete. The fertilized ovum is now referred to as a *zygote*. The fates of the middle and tail pieces of the penetrating spermatozoon are obscure, but it is possible that the centrioles and mitochondria they contain are used by the zygote.

The twenty-three chromosomes of the spermatozoon and the twenty-three chromosomes of the ovum replicate (i.e. manufacture copies of themselves) and form pairs of identical *chromatids* (*see Figure 16*). The members of the pairs separate and move to opposite poles of the cell, the movement being carried out by a structure called the *mitotic spindle*. This consists of an array of slender microtubules that arch across the cell between two centrioles. Initially the chromatid pairs form up around the equator of the spindle midway between the centrioles, and they are then moved apart. These nuclear changes are accompanied by a subdivision of the cytoplasm into two halves, and in this way the zygote undertakes its first division.

Cleavage

The first sequence of cell divisions is unusual in that no growth occurs. The single large zygote is converted into successively smaller cells whose total volume remains the same as that of the original zygote. There is no manufacture of new cytoplasm. This period is known as *cleavage,* and the outcome is a ball of cells still contained within the zona pellucida, as shown in Figure 5. As cleavage occurs, the developing embryo moves down the uterine tube towards the uterus, apparently

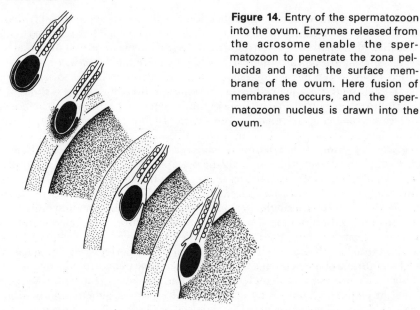

Figure 14. Entry of the spermatozoon into the ovum. Enzymes released from the acrosome enable the spermatozoon to penetrate the zona pellucida and reach the surface membrane of the ovum. Here fusion of membranes occurs, and the spermatozoon nucleus is drawn into the ovum.

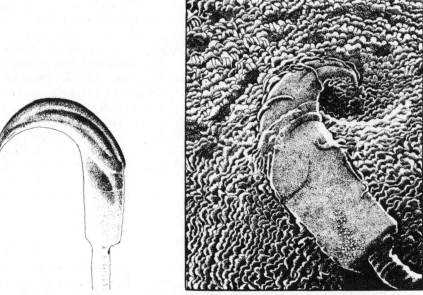

Figure 15. Hamster spermatozoa — drawings made from scanning electron micrographs. *Left:* A normal mature spermatozoon. *Right:* A spermatozoon probably undergoing fusion with the surface of the egg. [From Austin and Short (1972)]

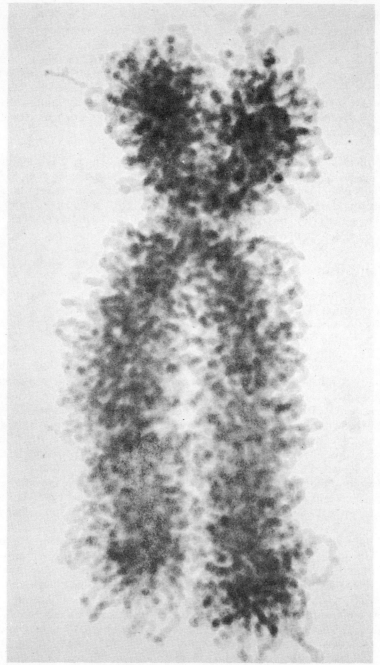

Figure 16. Human mitotic chromosome. The constricted portion represents the centromere connecting the two sister chromatids. [From DuPraw (1970)]

wafted by the action of the ciliated cells that line the tube. The metabolic requirements of the embryonic cells are met by the direct diffusion of nutrients into the cells from the fluid filling the tube.

At about the time the thirty-two-cell stage is reached, the zona pellucida disperses and the cells of the embryo are reorganized. The ball develops a central fluid-filled cavity, and the cells now form a hollow sphere called the *blastocyst*. Within this shell several cells migrate across the internal surface and congregate in a group known as the *inner cell mass*. Thus the embryo now consists of two distinct components:

1. A hollow shell of flattened cells whose subsequent role is to form the supporting tissues, in particular the placenta, for the embryo and foetus. This component is called the *trophoblast* to underline its role in nutrition during development.
2. The inner cell mass, which will form the embryo itself plus two chambers called the amnion and the yolk sac.

Implantation

Some 7 – 10 days after fertilization the embryo reaches the uterus and begins to invade its lining. This invasion is referred to as *implantation*. The mechanism of implantation and the subsequent relationship between the parasitic embryo and the surrounding maternal tissues are of considerable interest.

Implantation usually occurs in the wider upper end of the uterus. The blastocyst adheres to the epithelial lining and then penetrates it by enzymatic digestion. It sinks progressively deeper into the spongy oedematous tissue beneath, the *endometrium*, until it is totally submerged (*see Figure 6*). Since the endometrium is richly vascularized the outer shell of the blastocyst soon comes into contact with maternal blood, and this initiates a relationship that becomes increasingly refined as the organ of exchange – the placenta – is established.

Clearly, the cells of the blastocyst have a different genetic makeup from the cells of the mother's uterus, and therefore it might be expected that the mother would attempt to reject them, in the same way that a heterozygous graft of tissue is recognized and rejected by the immune system of the host body. However, rejection of the embryo does not occur, and this raises the question of why this should be.

This is still a controversial subject. Research has centred on the properties of the uterine wall and those of the conceptus. Is the uterus a 'privileged' site where the normal defence mechanisms of the body do not operate? Or are embryonic cells in some way different from other cells so that their alien nature cannot be recognized by the maternal cells? Is the immune response mechanism of the mother temporarily put out of action during pregnancy? Experiments have provided answers to some of these questions. For example, when heterozygous grafts are inserted into the uterine wall of experimental animals the grafts are rejected. This indicates that the uterus is not a privileged site: foreign cells are recognized and rejected as rapidly by the uterus as by other parts of the body. On the other hand, conceptuses can develop in a number of abnormal sites in the body provided that they have access to an adequate supply of maternal blood. Such abnormal implantation sometimes occurs spontaneously, and instead of implanting in the uterus the embryo may attach itself to the wall of the uterine tube, to the peritoneal lining of the abdomen, or to the mesentery of the intestines. This situation is called *ectopic pregnancy* and usually presents a serious clinical problem because of the risk of haemorrhage as the conceptus enlarges in sites that are not adapted for implantation. Alternatively, implantation can be produced experimentally in even more artificial sites, such as beneath the capsule of the kidney or within an eye from which the lens has been removed. Development without rejection still occurs until mechanical factors prevent further expansion. Thus it seems that the special properties that allow normal implantation and development without rejection are possessed by the conceptus rather than the uterus.

It has been more difficult to determine how this immunity is achieved. It has been known for about twenty years that early embryonic cells do not carry antigens on the outer membrane in the way that more mature cells do and that they therefore do not elicit an antibody reaction. However, specific antigens begin to appear towards the end of the embryonic period of organ building as cells differentiate, and from then on specific antisera can be produced experimentally. Antisera have the ability selectively to destroy developing organs or systems in the foetus, and it has been shown recently that some birth defects are caused in this way. Thus, although tolerance of the early embryo by the mother may be due to the lack of antigenic 'labels' on the cells, the subsequent months of pregnancy cannot be explained by the same mechanism. It is now thought that the maternal cells are prevented from coming into contact with the antigens produced by embryonic cells. This 'barrier' concept is gaining support from a number of studies, and it now appears either that the outermost cell layer of the trophoblast remains antigen-free or that a thin inorganic layer (mucoprotein) is deposited on the surface of the membrane and masks the trophoblastic cells.

Apart from the embryological significance of this interrelationship between the mother and the implanted conceptus, a greater understanding of the special qualities of the trophoblastic cells may provide ways of preventing or controlling the highly malignant form of cancer known as *chorioepithelioma*. This is caused by cells that break away from the placenta or by fragments of the placenta that remain in the uterus after childbirth. By virtue of their invasive and protected nature the isolated trophoblastic cells spread rapidly through the maternal body and produce tumours.

The embryo is now almost 2 weeks old and lies fully implanted in the uterine endometrium. Rapid changes begin to occur in both the trophoblast and the inner cell mass. Obviously the changes are interdependent and interlocked, but for the purposes of description it is best to consider them separately to avoid confusion. Thus development of the embryo and foetus will be described before an account is given of the changes that occur in the enveloping trophoblast. First, however, a note about the phenomenon of twinning.

Twinning

The nearly simultaneous birth of two babies to one mother is quite rare in humans and occurs in about 1 per cent of births, although in other species (e.g. the marmoset) it is more common and may be the general rule. There are two types of twins: *monozygotic,* in which both members develop from a single fertilized ovum and are thus identical, and *dizygotic* or fraternal twins, which develop from two different ova.

Monozygotic twins

About 25 per cent of twin births are monozygotic. There appear to be two stages of development at which the original conceptus can split to give rise to two babies. The first occurs very soon after fertilization, at the time when the blastomeres are still totipotent (i.e. potentially capable of giving rise to a complete individual, together with a complete set of supporting membranes, if they should become separated). Thus separation at the two-celled stage would give rise to two viable conceptuses, which would implant independently of each other and give rise to separate embryos, chorions, amnions, and placentas.

The second opportunity for monozygotic twinning arises after implantation and involves only the cells of the inner cell mass. This subdivision presumably occurs before the appearance of the primitive streak and implies that each of the groups of cells formed in this way retains sufficient potential to produce a complete embryonic body. As a consequence of this mode of twinning, the two individuals share a single chorion and placenta and may even lie within a single amniotic cavity.

On rare occasions the splitting of the inner cell mass may be incomplete, so that the organizing regions overlap or partially coincide, giving rise to *conjoined twins*. The degree of union ranges from a slight skin fusion, if the subdivision of the inner cell mass was almost complete, to a grotesque fusion of heads, limbs, trunks, and internal organs, if the separation was only partial (*see Figure 17*). Provided that the degree of fusion is quite mild, it is sometimes possible to separate conjoined twins surgically, but in the cases where a major organ such as the heart or liver crosses the point of fusion, separation is very hazardous. Incidentally, the commonly used term 'Siamese twins' for conjoined twins originates from the celebrated twins Chang and Eng Bunker, born in Thailand (Siam) in 1811. They died within three hours of each other at the age of 62 after fathering nineteen children between them. Rarely, one of the twins is deprived of an equal share of the placenta and grows comparatively slowly. It may also become increasingly dependent on an organ or system of the other twin, until it adopts in effect a parasitic role. The inequality in size in this situation may become very marked, and examples have been described in which the parasitic twin is no more than a tumour-like structure attached to the main twin or even incorporated into its body.

Since monozygotic twins have identical genetic constitutions they are similar in physical appearance, are of the same sex, and have identical blood groups. It is possible, however, for the hair and eyes to differ slightly in colour, and the degree of similarity between the twins becomes decreased if they are brought up in contrasting environments.

A Description of Normal Development

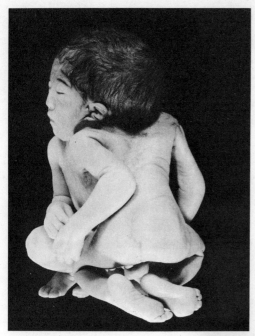

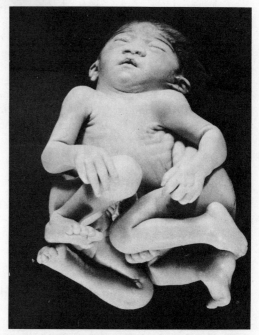

Figure 17. Effect of partial subdivision of the inner cell mass. The result in this case was a foetus with a single large head, four arms, and four legs. The lower part of the trunk is divided in two.

The uncanny physical resemblance of identical twins and their often-reported psychological closeness have led to many legends relating to their powers. In some cultures, the fear that twins might, when acting together, bring evil forces to bear on the community has prompted the killing of twins, as the following extract from a Nigerian newspaper reports:

> '... last week two sets of twins were strangled to death at a village near Abakaliki in the East-Central State. The reason given for the criminal act was that the twins are still regarded as an abomination and a curse to the entire village. The people of this village, therefore, need a more convincing proof that twins should be spared their lives. That proof should be based on the biological origin of twins.' (*Daily Times*, 1973)

Dizygotic twins

Dizygotic twins are produced when two ova are shed during the same ovarian cycle and each is fertilized by a spermatozoon, and thus they are no more alike than babies produced in separate pregnancies. About 75 per cent of twin births are the result of this process. The ova may come from one ovary following rupture of two follicles or a follicle containing two oocytes, or both ovaries may contribute one ovum each. The conceptuses implant separately, and usually there is no sharing of extraembryonic tissues, although the placentas may expand towards each other and sometimes appear to fuse. Very rarely there is mixing of the two foetal circulations in the region where the placentas are in contact, and this can produce complications if one twin is male and the other female since the male sex hormone produced in the male foetus will be carried into the female and may produce masculinization of her developing reproductive system.

There are marked ethnic differences in the incidence of dizygotic twins. The incidence is low in Japan, being just over 1:1 000 births, while in Nigeria the rate is 40:1 000 births. The incidence also seems to be linked with maternal factors: tall older multiparous women from upper social strata are more likely to have dizygotic twins than other mothers.

Triplets, quadruplets, and above

Higher orders of multiple births are very rare. In a population in which the incidence of twin births is 1:100, the incidence of triplets would be only 1:10 000 births and of quadruplets only 1:1 million births — a hundredfold increase in rarity at each step. The risks to the mother and babies increase in direct proportion to the number of foetuses, and prematurity with its attendant hazards is one of the major consequences.

The foetuses may all develop from a single zygote, in which case they are monozygotic, or they may develop from separate ova and thus be trizygotic, etc. Alternatively, both processes may occur, and some foetuses would then be identical while the remainder would have different origins.

The use of hormone therapy in the treatment of infertile women has resulted in a number of cases of high-order multiple pregnancies, with reports of six, seven, or even eight foetuses. The treatment consists of injections of follicle-stimulating hormone followed by chorionic gonadotrophic hormone, and if the dose given is too high a number of ova mature and are ovulated.

After this diversion, let us return to the account of embryonic development.

The Embryonic Period

Formation of the germ layers

At first, the inner cell mass consists of an irregular group of cells of various sizes, but soon after implantation a single layer of cells becomes organized on its internal, or luminal, surface. This layer is called the *endoderm* and constitutes one of the *germ layers* of the embryo, of which there are ultimately three. The remaining cells of the inner cell mass form another germ layer: the *ectoderm*. Spaces appear and coalesce within the clump of ectodermal cells until a primitive *amniotic cavity* is formed. The ectodermal cells remaining in contact with the endodermal layer become organized into a coherent layer of columnar cells, and the embryo proper now consists of a bilaminar plate of cells crossing the lumen of the hollow trophoblast. The bilaminar embryo is approximately circular in shape and is referred to as the *embryonic disc (see Figure 6)*.

The endodermal cells proliferate, and the endoderm expands around the inside of the trophoblastic cavity to form a bag-shaped *yolk sac*. This is perhaps a misleading term when applied to mammalian embryos since the yolk sac does not contain yolk, but the label is retained to indicate that the endodermal sac is analogous to the yolk-filled structures in embryos of nonmammalian vertebrates that depend on stores of yolk for nourishment during development.

Another endodermal structure begins to develop at this time. It originates as a small diverticulum of endoderm near the periphery of the embryonic disc. This later becomes the *allantois,* a pouch-like structure that in some species acts as a temporary storage chamber for nitrogenous wastes during development, but in mammalian embryos plays a less significant role and finally contributes only a small part of the bladder.

The trophoblastic shell grows rapidly, and the embryonic disc with the attached amnion and yolk sac occupy only a small part of the central cavity. Cells derived from the inner surface of the trophoblast form a loosely arranged tissue called the *extraembryonic mesoderm* in the intervening region.

Soon the extraembryonic mesoderm changes from its loose reticular arrangement and becomes organized into two fairly coherent layers: one around the inside of the trophoblast and the other around the outside of the amnion and yolk sac. A bridge of compact mesoderm joins the embryo to the surrounding trophoblast, but apart from this *body stalk* there are few cells in the intervening fluid-filled space separating the two. This cavity is referred to as the *extraembryonic coelom* (from the Greek: *coelom* = cavity). The endodermal allantois is incorporated into the mesoderm of the body stalk.

The two-layered embryonic disc at first shows no indication of polarization into head and tail regions, but then on the 15th day a linear thickening develops in the ectoderm extending from approximately the centre of the disc to the periphery (*see Figure 7*). This thickening is produced by the confrontation of two opposing streams of ectodermal cells. The motivation for the streaming is not clear, but it is supported by very active proliferation of ectodermal cells around the periphery of the embryonic disc.

The linear ectodermal thickening is called the *primitive streak*, and it has three significant functions.

1. It defines the midline of the embryo.
2. It establishes the polarity of the embryo by extending from the centre of the disc to what will become the caudal end of the embryo.
3. It generates the third embryonic germ layer, the *embryonic mesoderm*, by a process of invagination of ectodermal cells into the region between the ectoderm and endoderm.

Formation of the embryonic mesoderm has not been studied extensively in human embryos because of the lack of suitable material, but it seems probable that the process is very similar to that known to occur in other species. The following account is based mainly on studies of chick embryos.

The primitive streak appears in cross-section as a thickened region marked on the surface by a shallow groove with slightly raised edges, as shown in *Figure 18*. The ectodermal cells forming the floor of the groove are very elongated perpendicularly in relation to the surface, and the nucleus of each cell is placed at the deep end of the cell close to the endodermal layer. This positioning of the nucleus gives the cell an asymmetric shape, with the large cell body deep to the surface and a narrow process extending to the surface. Electron microscopy has revealed that the elongated processes of invaginating cells contain elongated bundles of longitudinal microtubules. These microtubules are identical in appearance and dimensions to the microtubules forming the mitotic spindle of dividing cells. The

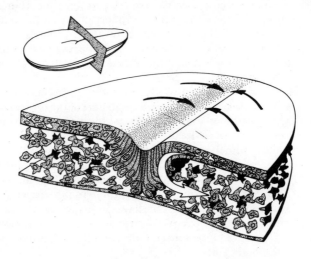

Figure 18. The primitive streak. Ectodermal cells stream towards the midline in the caudal region of the embryonic disc and invaginate to form mesodermal cells. *Upper left:* Diagram showing the plane of section.

spindle microtubules provide the power to move daughter chromosomes to opposite poles of the dividing cell, and so it seems reasonable to suggest that the bundles of orientated microtubules that appear in certain groups of cells during embryogenesis may also be involved in producing movements of some sort. Generally it has been found that the appearance of microtubules is associated with movement of the nucleus towards one end of the cell together with a general elongation of the cell.

Thus, in the case of the ectodermal cells forming the primitive streak, it is thought that the microtubules produce marked elongation of the cells, pushing the nucleus and cell body to a deeper position from which the cell can sever its connection with the surface and then migrate away between the ectoderm and endoderm. Invagination occurs most actively at the end of the primitive streak positioned near the centre of the embryonic disc. This region is commonly referred to as Hensen's node. The mesodermal cells produced are irregular and amoeboid in shape, and they move away from their point of origin, gradually building a loose meshwork of cells between the original two germ layers.

Structure of the three germ layers

Before describing how this 'embryonic sandwich' is converted into a complex organism, it may be helpful to note the structural differences that exist between the three germ layers. Briefly, the ectoderm is a fairly substantial and coherent layer, the mesoderm is a more spongy meshwork of cells, at least at first, while the endoderm is initially a thin continuous membrane of very flattened cells, as shown in *Figure 19*.

The ectoderm generally contains two interlocking layers of cells organized so that the internal and external surfaces are smooth. There are intercellular spaces within the ectoderm, but where the cells touch each other there are special junctions that strengthen the contact. There is a basal lamina close to the internal surface of the ectoderm. A basal lamina consists of a fairly diffuse layer of fibrous proteins lying parallel to the surface of a sheet of cells. Its function is not known but may be of significance in some embryological processes, such as induction. Perhaps as a consequence of its tightly knit organization the ectoderm is often involved in gross folding movements to produce tubular or vesicular structures in the embryo.

In contrast with the ectoderm the mesoderm consists of loosely arranged, irregular cells. The cells tend to be migratory in the early embryo, moving between the ectoderm and endoderm and often showing long processes extending away from the cell body to make contact with nearby cells. It is unusual to see areas of strengthened contact between cells, although special close junctions occur and are thought to facilitate the exchange of molecules between cells. In the mesoderm structures are established by activities such as migration, aggregation of cells, or reorganization of the cellular lacework into thin epithelia bordering spaces such as body cavities and future blood vessels.

The endoderm consists of flat plate-like cells that form a layer. It is similar to the ectoderm in that it possesses a basal lamina internally but different in that it is thin and only one cell thick at this stage. The endoderm appears to play a fairly passive role at first, and it seems to be moulded into a gut tube largely by the activity of the other germ layers.

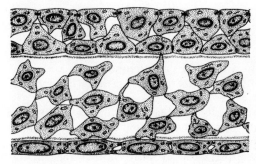

Figure 19. The three embryonic germ layers — an idealized drawing. The ectoderm, shown at the top, has a two-layered organized appearance and is lined internally by a basal lamina. The middle layer, or mesoderm, is composed of an open meshwork of irregular cells. The third layer, or endoderm, is at first a thin sheet of flattened cells, again lined by a basal lamina. Specialized intercellular junctions exist between cells in each layer and between cells of adjacent layers.

Development from the germ layers

In two regions of the embryonic disc the ectoderm and endoderm remain firmly attached to each other, and at no stage do the migrating mesodermal cells penetrate between them. One of these regions occurs at the future head end, the other at the tail end, and both lie on the midline. The more cranial of the two is called the *buccopharyngeal membrane*, which will become a significant feature in relation to the developing mouth, while the more caudal of the regions is the *cloacal membrane*, which plays a part in development of the hindgut and the urinary and genital systems (*see Figure 20*).

Elsewhere, the mesodermal cells begin to aggregate and condense to form structures. The first to appear is known as the *notochord:* a median rod-like collection of cells extending cranially from Hensen's node. In less complex animals the notochord is retained as a stiffening rod in the adult, but in mammalian embryos it is superseded in this function by the developing vertebral column, although probably the notochord plays a supporting role during early stages. Remnants can be seen in the adult: the central nucleus pulposus of each intervertebral disc is thought to be derived from the notochord. Classical descriptions of the notochord in human embryos often refer to a central canal running along its length, opening into the amniotic cavity at Hensen's node and into the yolk sac cavity at the other end. However, more recent studies — especially those using the electron microscope — have shown the notochord to be built up of wedge-shaped cells that in cross-section look like the segments of a halved orange, and there is no evidence of a central canal (*see Figure 21*). It is possible, therefore, that the canal appearing in the earlier studies was a shrinkage artefact resulting from inappropriate preparation for microscopy.

Formation of the notochord is followed by several important changes. The ectoderm lying over the notochord becomes thickened as a result of elongation of the component cells (*see Figure 22*). This is the first step in the formation of the *neural plate*, which is the precursor of the central nervous system (i.e. the brain and spinal cord). The neural plate is oval in outline when viewed from the amniotic aspect and soon develops a midline furrow extending throughout its length. The furrow deepens, and concurrently the sides of the groove rise until eventually they arch over the floor of the groove and fuse, thus forming a tube of ectodermal cells. Details of this process of *neurulation* will be given in the subsection dealing specifically with the development of the nervous system, but it is interesting to

A Description of Normal Development 31

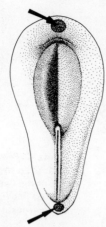

Figure 20. The positions of the buccopharyngeal membrane (top) and the cloacal membrane (bottom), shown diagrammatically.

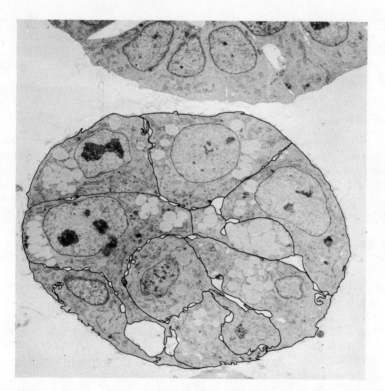

Figure 21. Ultrastructural appearance of the notochord in transverse section. The component cells are generally wedge-shaped with their apices directed centrally. This specimen came from a chick embryo that had been irradiated with 300 rads of X-rays twenty-four hours before. Although other structures were malformed, the notochord has retained its normal organization, showing it to be resistant to X-rays once it has been established. For clarity, cell boundaries have been outlined. (x 4800).

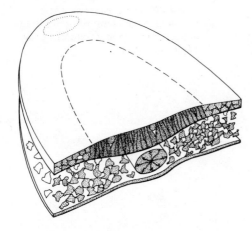

Figure 22. Formation of the neural plate in relation to the notochord. The outline of the neural plate is marked by a dashed line. In front of it lies the buccopharyngeal membrane.

note that microtubules play a part in this event as they do in other morphogenetic movements and that *induction* of the ectodermal cells must occur if the neural plate is to develop normally. The phenomenon of induction was mentioned in the introductory section, but in view of its significance it will be worthwhile to consider it in more detail at this point.

Induction

During development, some regions of the embryo strongly influence the development of adjacent regions. The dominant region is referred to as an *organizer* or *inducer,* and it elicits changes in the responding tissue that would not occur otherwise. Although the mechanism of this process is not clearly understood, there is experimental evidence to suggest that the effect is brought about by chemical transmission. Various substances such as protein, nucleoprotein, glycogen, and steroids have been put forward as possible inducers, but interpretation of experimental work became more difficult when it was found that many nonphysiological treatments can also produce changes characteristic of inductive effects. To explain this, hypotheses have been proposed that give inducers the role of nonspecific 'triggers' rather than that of specific information-carrying molecules as it was originally assumed.

Attention was first drawn to the significance of induction by Spemann, who showed by a series of careful experiments that development of the early amphibian embryo is controlled by a specific region: the dorsal lip of the blastopore. This region lies at the rim of an opening, the blastopore, through which surface cells invaginate to an internal position in the embryo, which, unlike mammalian embryonic discs, is hollow and spherical at this early stage. He called this special region the *'primary organizer'* and showed that it stimulated formation of the neural plate in the ectoderm nearby. Later studies identified comparable regions in embryos of a number of other species. In the chick embryo, for example, it is thought that Hensen's node is analogous to the dorsal lip of amphibian embryos. It has been argued, however, that organization is already apparent in the blastoderm before the node region becomes active, so the use of the term 'primary organizer' in this case may not be appropriate.

A Description of Normal Development 33

Inductions occur throughout organogenesis. For example, it has been shown experimentally that the neural tube, the lens of the eye, and the nephrons of the kidneys all depend on inductive interactions. The responding tissues are only sensitive to the inducing stimulus for a brief period, however. If for some reason induction does not occur, the tissues now begin to differentiate in an abnormal way. Thus the lens of the eye does not develop unless the ectoderm is induced to do so by the optic vesicle.

The failure so far to identify specific inducing molecules has led to concepts of *developmental fields* — the word 'fields' in this context having the same connotations associated with its use in physics (e.g. in magnetic field or gravitational field). One theory is that such developmental fields consist of overlapping physiological gradients of nonspecific substances, such as metabolites or nutrients, whose pattern of distribution or concentration in some way influences the development of different embryonic regions. This concept will be explored in more detail on page 154.

Induction in the classical sense together with physiological gradients are unlikely to be the only intercellular communications playing significant roles in development. Interchange probably also occurs at specialized *intercellular junctions* that are distributed throughout the embryo during development. Two types of junction in particular are of interest: close junctions, in which the outer cell membranes of adjacent cells are separated by a space of less than 10 nanometres, and tight junctions, in which the cell membranes are in direct contact. These specialized junctions tend to be very limited in extent and can only be studied by electron microscopy. Small molecules can pass quite freely across such junctions, and even molecules up to a molecular weight of 69 000 have been shown to cross successfully, so it is quite possible that there is a mechanism of chemical communication between cells linked in this way.

Embryonic development continued

The neural plate is established over quite an extended period of development. It first appears on the 18th day, and closure is completed on the 26th day. The future brain region appears and closes first, and the remaining parts of the neural plate are added to the caudal margin of the first-formed region. This is made possible by a steady caudal movement or *regression* of the primitive streak, which is still a site of active invagination of cells to form mesoderm. As Hensen's node regresses, cells are continuously added to the notochord so that this grows in length in association with the overlying neural plate. The embryo has by this time lost its circular shape and become more elongated. (*See Figures 7 to 9.*)

On each side of the closing neural plate the mesodermal cells begin to aggregate and form regular blocks called *somites*. They increase in number as the neural plate lengthens, and two parallel columns of somites are formed. The somites give the embryo a segmented appearance, as shown in Figure 8. The first somite appears on the 20th day, and by the 35th day the final complement of forty-three pairs of somites is present. With further development the somites lose their clearly defined segmental arrangement, but the muscular, skeletal, and dermal tissues derived from them still show evidence of this segmental origin.

The mesoderm situated lateral to the columns of somites also becomes modified. Towards the periphery of the disc it becomes split into two layers, one layer remaining in association with the ectoderm and the other with the endoderm. (*See Figure 23.*) With reference to their future roles in development these layers are named respectively *somatic mesoderm* and *visceral mesoderm*, since the ectoderm and somatic mesoderm will form much of the body wall while the endoderm and visceral mesoderm will form most of the viscera and associated structures. The fluid-filled cleft that separates the somatic mesoderm from the visceral mesoderm is called the *intraembryonic coelom* and communicates with the extraembryonic coelom at the lateral margins of the embryonic disc.

A solid column of mesoderm links the medial somites with the more laterally placed somatic and visceral mesoderm. This is the *intermediate mesoderm* within which the urinary and genital systems develop.

As the neural plate closes the basic shape of the embryo begins to change. Prior to neurulation the embryonic disc is a flat membranous structure, but with formation of the neural tube (especially the large brain region) and its subsequent rapid growth the embryo begins to mushroom into the amniotic cavity. The central regions become elevated in comparison with the periphery of the disc. An idea of these changes can be gained by modelling the situation with a sheet of cloth. The cloth is placed on a flat surface to represent the embryonic disc, and the arm is then slipped underneath with the fist clenched to represent the brain region and future spinal cord. If the arm is now lifted from the surface and pushed

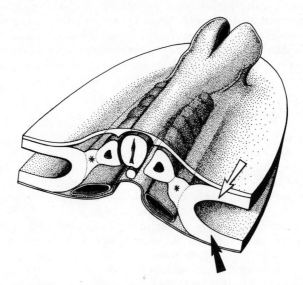

Figure 23. Subdivisions of the embryonic mesoderm. The somitic mesoderm forms a series of segmental blocks on each side of the neural tube. Just lateral to these lies a column of intermediate mesoderm, marked here with an asterisk, in which the urogenital system develops. Towards the periphery of the embryonic disc the mesoderm is split into two layers: the upper layer, marked by a white arrow, is the somatic mesoderm; and the lower layer, marked by a black arrow, is the visceral mesoderm. Lying between these two layers is a fluid-filled space called the intraembryonic coelom.

forward to represent the growth of the neural tube, an impression of the changes that occur in the embryo may be gained from the new appearance of the piece of material.

Eventually the peripheral regions of the original embryonic disc become gathered in beneath the ventral aspect of the embryo. As a result, the endodermal layer becomes folded to form the primitive gut, the embryo as a whole takes on a much more tubular appearance than before, and its attachment to the inside of the trophoblastic shell consists now of a narrow stalk covered with amniotic ectoderm.

In the remainder of the embryonic period all the major organ systems become established, so to simplify the account each system will now be considered in turn.

The Nervous System

Development of the neural tube

After formation of the notochord the neural plate thickens and folds to form the neural tube. The mechanism of closure has been closely studied in amphibians and chick embryos, and it now seems that two intracellular organelles play an important part in this process. These are *microtubules,* which have already been mentioned in the context of the primitive streak, and *microfilaments,* which are found in the dorsal region of the neural plate cells and are probably contractile. Blockage of the formation of either microtubules or microfilaments by the administration of certain drugs results in a failure of the neural plate to close. It is suggested that these organelles co-operate in the following way: the microtubules produce elongation of the cells of the neural plate, also pushing the nuclei to a deeper position within the cells, while the microfilaments contract at the amniotic border like purse-strings, causing each cell to become wedge-shaped in cross-section (*see Figure 24*). When this change occurs throughout the neural plate it begins to buckle. Since there is an intimate relationship between the stiffening notochord and the midline of the neural plate, this buckling takes the form of a longitudinal groove that gradually deepens. This process is probably aided by the changes in shape of the nearby somites, and there may also be a contribution by the ectoderm adjacent to the neural plate, pushing the neural folds together in the midline.

During fusion of the neural folds a wedge-shaped column of ectodermal cells becomes apparent along the line of contact (*see Figure 25*). This wedge becomes split into dorsal and ventral components. The dorsal portion divides longitudinally, and the cells in each half migrate in a lateral and then ventral direction around the outer surface of the neural tube. These cells constitute the *neural crest,* and they have eventually very diverse fates, becoming involved, for example, in formation of parts of the peripheral nervous system, the adrenal medulla, and pigment cells beneath the skin. The ventral portion of the wedge is retained as the roof of the neural tube and rapidly becomes indistinguishable from the rest of the tube. After fusion of the neural folds the ectoderm that was originally peripheral to the neural plate now fuses in the midline to form a continuous cell layer over the neural tube.

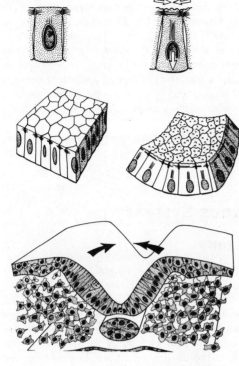

Figure 24. The mechanism of neurulation. The changes in shape (believed to be brought about by microfilaments and microtubules) are shown for individual cells *(top)*, plates of cells *(middle)*, and the neural plate with associated notochord *(bottom)*.

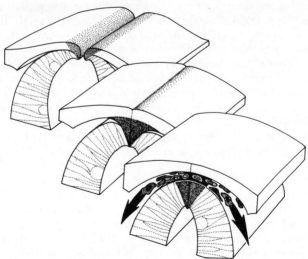

Figure 25. Cells of the neural crest. At the time of fusion of the neural folds *(upper left)*, the neural crest cells are situated in the junctional zone between the neural plate and the more laterally placed ectoderm. After fusion *(middle)*, they form a wedge-shaped column. They are then *(lower right)* split into a dorsal migratory group and a ventral group that completes the roof of the neural tube.

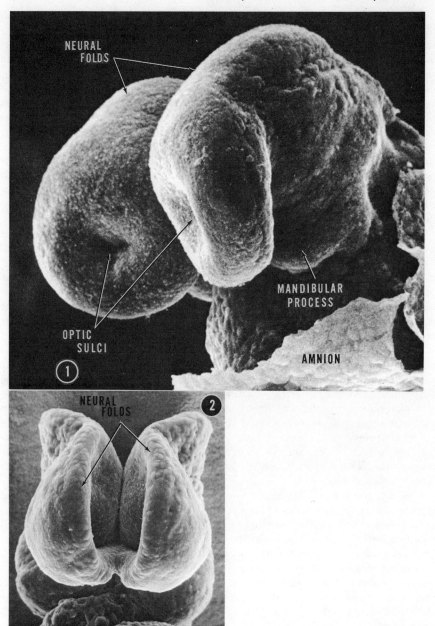

Figure 26. The developing brain region of a hamster embryo − scanning electron micrographs. (1) In this 7¾-day embryo the neural folds are wide open. (2) Twenty-four hours later the neural folds are beginning to approximate. [Micrographs from Waterman, R. E. (1972) 'Use of the scanning electron microscope for observation of vertebrate embryos,' *Developmental Biology*, **27**, 276]

Closure of the neural tube begins in the future midbrain region and extends both cranially and caudally from this point (*see Figure 26*). The cranial end of the tube closes completely by the 24th day, and the caudal region closes a little later on the 26th day. Closure of the embryonic brain is followed by a period of rapid expansion, largely as a result of the accumulation of fluid within that region of the neural tube. It has been noted that in chick embryos a higher pressure of fluid exists within the enlarging brain than in the amniotic cavity. This observation poses an intriguing problem since the lumen of the neural tube communicates with the amniotic cavity through the unclosed caudal region of the neural plate for the first stage of this period of expansion, and it is difficult to explain how this raised pressure is maintained. Interestingly, immediately after closure of the neural tube the border surrounding the central lumen becomes ciliated, so it is possible that the cilia can actively produce the pressure differential. Alternatively, the lining of the neural tube may secrete fluid more rapidly than it can escape into the amniotic cavity, thus leading to an accumulation.

The neural tube is not formed entirely by a process of closure of a neural plate. When closure of the plate is complete, further development of the caudal neural tube proceeds for a limited time by the activity of a structure called the *end bud*. Little is known about this structure, but it extends caudally from the termination of the primary neural tube and consists of a solid midline cord of cells in which a lumen develops. The lumen is a blind continuation of the central canal of the original tube, and it extends, as the end bud elongates, by a process of cellular reorganization rather than degeneration.

The neural tube now comprises a dilated brain region and an elongated caudal portion with a smaller diameter, which will become the spinal cord. The wall of the tube is formed of a *pseudostratified epithelium*, which is a shorthand way of saying that the cells are organized in a close-knit layer, or epithelium, with many − although not all − of the cells stretching from one surface to the other surface. In this case, many of the cells extend from the internal (luminal) surface to the external surface of the neural tube. The remaining cells all border onto the luminal surface but do not extend across the entire thickness of the wall of the neural tube. These shorter cells are involved in the cycle of changes that occur during cell division.

Cell division occurs only at the luminal border (*see Figure 27*). The elongated cells round up to this border to divide, probably because the cell-to-cell contacts are particularly strengthened close to the lumen. The total time taken by a cell when it divides is eight hours. During prophase, the nucleus moves towards the luminal surface and the external process of the cell is drawn after it, the cell becoming almost spherical. During metaphase, the chromatids separate in a plane parallel to the luminal surface. As the daughter cells separate a process regrows from each towards the external limiting layer, and both cells become elongated during telophase until they have regained the configuration of their predecessor. The luminal border of the neural tube has a very crowded appearance with numerous cell processes packed side by side, and electron micrographs give the impression that there is little space available for cell division. Dividing cells cause a noticeable distortion of the neighbouring cell processes, and it is possible that this degree of crowding may in some way control the rate of cell division.

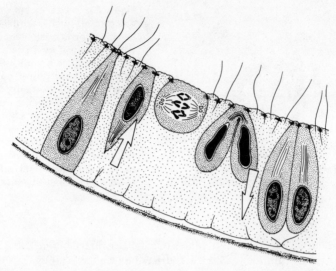

Figure 27. Cell division in the neural tube. From left to right the following stages are shown: interphase, prophase, metaphase (at the luminal border), early telophase, and late telophase.

Development of the spinal cord

The repeated division of cells in the wall of the caudal part of the neural tube produces a steady growth in size, which is then followed by a period of reorganization. Primitive nerve cells — *neurocytes* — are formed, and these will not divide again. They migrate centrifugally and form a zone called the *mantle layer* around the neuroepithelium. The mantle layer later forms the grey matter of the spinal cord — the core region that contains predominantly the cell bodies of nerve cells.

The neurocytes now develop long fibres that grow peripherally and form a layer external to the mantle layer called the *marginal layer*. At a later stage, beginning in the 4th month and continuing after birth, these fibres become insulated with myelin sheaths and give the marginal layer a white appearance, and so this layer is referred to as the white matter of the spinal cord. The spinal cord is not composed entirely of nerve cells and their processes, however, since in addition three types of essential supporting cells — *glial cells* — are developed. Two types of glia are produced within the neuroepithelium, and these migrate peripherally into the mantle and marginal layers. The third type is formed from mesodermal cells, which migrate into the developing spinal cord.

The walls of the neural tube become much thickened while the roof and floor plates remain comparatively thin (*see Figure 28*). The central lumen becomes narrowed, forming a dorsoventrally elongated cleft bordered by the neuroepithelium, which, after its period of intense proliferative activity, is converted into a layer of more quiescent *ependymal cells*. Two longitudinal columns of grey matter can soon be discerned in each of the thickened lateral walls. The ventral column is called the *basal plate,* and the dorsal column is called the *alar*

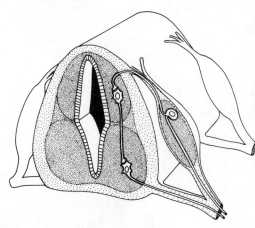

Figure 28. The developing spinal cord. The central lumen is bordered by a thin layer of ependymal cells, external to which lies the grey matter of the cord. The grey matter is subdivided into alar (upper) and basal (lower) components. Overlying this is the marginal layer that later becomes the white matter of the spinal cord. The formation of spinal nerves by the fusion of dorsal and ventral roots is also shown, together with the positioning and connections of nerve cells in the dorsal root ganglion, the alar plate, and the basal plate.

plate. The neurocytes of the basal plates become organized before those of the alar plates. Initially, a number of short dendrites (specialized cell processes that will receive impulses from other nerve cells) develop at one end of each cell and a single long axon develops from the other end. The differentiating cells become very large and are referred to as *multipolar neurons* because of their many-branched configuration. The axons of the basal plate neurons leave in bundles that are segmentally arranged like the somites. Each bundle is known as the *ventral root* of a particular spinal nerve, and its function is to carry impulses to muscle cells derived from the same level; i.e. it has a *motor function*.

The sensory components of the spinal nerves are derived from part of the neural crest, so the cell bodies are situated outside the central nervous system. Clusters of neural crest cells collect on each side of the spinal cord, and some of the cells differentiate into neurocytes. Each neurocyte develops two processes: one grows peripherally and develops a receptor organ at its tip, while the other grows medially towards the spinal cord and enters its dorsal surface. Having entered the spinal cord some of these central processes end in the grey matter of the alar plate, but others turn cranially in the marginal layer and pass directly to the brain. The bundles of central processes entering the spinal cord are called the *dorsal roots* of the spinal nerves, and the segmental groups of cell bodies in the dorsal roots are called the *dorsal root ganglia*. The peripheral processes of each dorsal root ganglion join the ventral root of the same level at a point just lateral to the ganglion to form the trunk of the mixed (motor and sensory) spinal nerve.

In response to the input of fibres by the sensory neurocytes, the neurons of the alar plate also develop processes that enter the marginal layer and either pass up or down the spinal cord to link segments or cross to the other side of the cord to co-ordinate the activity of both sides.

During the embryonic period the developing spinal cord completely fills the canal formed by the vertebrae. In the foetal period, however, the vertebral column grows more rapidly than the spinal cord. Since the spinal cord is firmly attached to the brain, which is itself anchored within the cranial cavity, this differential growth is compensated for by a gradual recession of the caudal tip of the spinal cord up the vertebral canal to a more cranial position. By the time of birth the spinal cord extends caudally to the level of the third lumbar vertebra, and in the adult it reaches only to the lower border of the first lumbar vertebra. As a consequence of this recession, the spinal nerves that originally passed laterally at right angles to the cord now have to pass caudally within the vertebral canal to reach the appropriate intervertebral foramina through which to exit. The degree of this caudal deviation is most marked for the caudal spinal nerves where the relative recession has been greatest, and the vertebral canal beyond the receding tip of the spinal cord is thus partially filled by a bundle of spinal nerves that, because of its resemblance to a horse's tail, is called the *cauda equina*.

Development of the brain

After closure of the neural tube in the brain region, three compartments or primary brain vesicles become apparent: the *forebrain vesicle*, the *midbrain vesicle*, and the *hindbrain vesicle*. (*See Figure 29.*) In the 5th week the forebrain and hindbrain vesicles become subdivided. The forebrain vesicle gives rise to two *cerebral vesicles*, which together are known as the *telencephalon*, and a central region connecting them called the *diencephalon*. The hindbrain vesicle becomes divided into a cranial portion, which will give rise to the future *pons* and *cerebellum*, and a caudal portion, which will become the *medulla oblongata*. As a result of these changes the original central lumen of the brain region is converted into a number of communicating chambers, which are the forerunners of the *ventricular system* of the brain (*see Figure 30*). The cavity in each cerebral hemisphere becomes a *lateral ventricle*, the cavity in the diencephalon becomes the *third ventricle*, and the cavity of the hindbrain vesicle becomes the *fourth ventricle*. The cavity of the midbrain vesicle does not enlarge in proportion with the growth of the rest of the brain, and it remains as a narrow canal linking the third ventricle with the fourth ventricle. This canal is known as the *cerebral aqueduct*. The fourth ventricle is continuous caudally with the central canal of the spinal cord.

Both the ventricular system and the central spinal canal are lined by secretory ependymal cells, which are the first source of the watery *cerebrospinal fluid* (CSF). Later, during the foetal period, CSF is produced as a modified filtrate of the blood serum at special sites in the ventricles called *choroid plexuses*, where thin regions of the brain wall are invaginated by tufts of blood vessels. At first the CSF remains trapped within the central nervous system, but during the 4th month of development three perforations appear in the thin membranous roof of the fourth ventricle. CSF can now escape from the ventricular system into a spongy tissue lying between the brain and the developing skull. This tissue (the *arachnoid mater*) is the intermediate layer of the three meninges that cover and protect the central nervous system. The innermost layer (the *pia mater*) is very thin and delicate (the name means gentle mother) and follows all the intricate surface contours of the brain and spinal cord. The outermost layer (the *dura mater* − literally 'tough mother') is tough and densely fibrous. This is tightly attached to the inside of the skull, but it also provides a number of fibrous septa that partially subdivide the cranial cavity and prevent undue movement of the brain. The arachnoid mater forms a coherent lining to the dura mater and sends strands of cells across the irregular space between the dural and pial layers. It is from these web-like strands of cells that the arachnoid takes its name (derived from the Greek for cobweb).

The CSF that enters the arachnoid is eventually absorbed into the blood carried by venous channels leaving the cranial cavity. Thus a *circulation of*

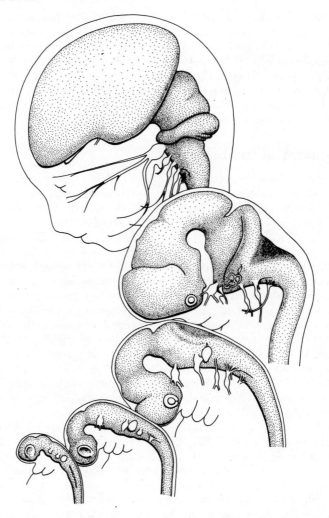

Figure 29. Development of the brain. The stages shown are: 3½ weeks *(bottom)*, 4 weeks, 5½ weeks, 7 weeks, and 11 weeks *(top)*.

cerebrospinal fluid is established: the fluid produced within the brain passes out through the perforations in the roof of the hindbrain, enters the arachnoid layer, and from there is absorbed into the bloodstream. An obstruction at any point in this pathway can cause an accumulation of CSF, a condition known as *hydrocephalus*. For example, a tumour in the midbrain close to the cerebral aqueduct may constrict it sufficiently to cause accumulation of fluid in the two lateral ventricles and the third ventricle; a failure of the roof of the fourth ventricle to perforate may cause enlargement of all the ventricles; malabsorption of CSF into the venous bloodstream causes enlargement of the arachnoid spaces. All these variants of hydrocephalus can cause serious damage to the brain.

A Description of Normal Development 43

Figure 30. The ventricular system of the brain and the flow of cerebrospinal fluid. *Left:* The ventricles and their communications are shown in diagramatic form. The two lateral ventricles drain into the median third ventricle, which is connected with the fourth ventricle by the narrow cerebral aqueduct. The fourth ventricle is itself continuous with the central canal of the spinal cord. The arrows indicate the formation and flow of cerebrospinal fluid, which finally exits from the ventricular system through the roof of the fourth ventricle. *Right:* The actual shape of the ventricular system is more complex than suggested above, but the same pattern of flow exists.

Initially, the embryonic disc is flat and the neural plate closes to form a fairly straight tube. Later the great enlargement of the brain region causes a longitudinal curvature, convex dorsally, to develop in the neural tube. Soon a more pronounced curvature of the neural tube occurs at the junction of the hindbrain vesicle with the future spinal cord. This is known as the *cervical flexure*. A series of flexures then develops throughout the brain region and is accompanied by a pronounced flexure of the caudal end of the neural tube – the so-called *tail flexure*. Thus, by the end of the embryonic period the neural tube has a convoluted shape.

Final organization of the brain

The internal changes that occur during development of the brain are extremely complex and can only be followed satisfactorily if an attempt is made to understand the final organization of the brain. In an introductory account such as this it is clearly inappropriate to go into great detail, but it may be of some use to note a few key points here as guidelines in case the reader is tempted to turn to a more comprehensive text.

The organization of the neural tissue into grey and white matter and the subdivision of the grey matter into alar (sensory) and basal (motor) plates occur in the brain region as they do in the spinal cord, but the distinctions are less clear-cut as a result of the superimposition of more complex patterns of organization. The degree of modification increases from the hindbrain towards the forebrain. The cerebellum, a large neural structure that will later play an important part in co-ordinating muscular activity, is developed from the dorsal parts of the alar plates in the hindbrain region. The diencephalon, one of the three components of the original forebrain vesicle, gives rise to many important structures: e.g. the optic vesicles (*see* p. 56), the pineal body, part of the pituitary gland, the thalami, and the hypothalamus. The pituitary gland is developed from two sources: an ectodermal upgrowth from the primitive mouth cavity and a downgrowth from the diencephalon. This two-part origin explains the marked contrast between the composition of the anterior lobe, which is derived from the ectoderm, and that of the posterior lobe, which is derived from the brain.

The two cerebral hemispheres are the other major structures developed from the original forebrain vesicle. They grow and expand rapidly, at first forward to form the *frontal lobes,* then laterally and upwards to form the *parietal lobes,* and finally posteriorly and inferiorly to produce the *occipital* and *temporal lobes.* As a result of this great expansion the hemispheres overshadow and partially bury the midbrain and diencephalon. The outer layer of the hemispheres is composed of grey matter and is known as the *cerebral cortex.* In order to accommodate a very large community of neural cells, the cortex becomes very convoluted during later development and the hemispheres take on their characteristic wrinkled appearance, in some ways resembling the kernel of a walnut.

These changes in external appearance are accompanied by intricate internal development of fibre pathways and cell-to-cell relationships. Even at the time of birth only certain vital centres of the brain are functional, and maturation of the physical basis of the brain continues for many years after birth.

The Heart and Major Blood Vessels

Development of the blood and preliminary capillaries

During the 3rd week of development the embryo can no longer satisfy its nutritional requirements by the simple diffusion of nutrients through its structure. To cope with the increasing demands of the rapidly growing embryo, a blood circulatory system develops to transport materials to and from the trophoblastic shell so that exchanges can occur more effectively between the maternal supporting tissues and the growing embryonic tissues. This system is the first to become functional in the embryo.

Clusters of mesodermal cells in the wall of the yolk sac, in the body stalk, and in the lining of the trophoblast begin to differentiate into primitive blood cells. The cells around these *blood islands* become flattened and form a continuous membrane, or endothelial lining, around each. Fluid that accumulates between the blood cells forms the supporting *plasma.* As the blood islands increase in size and number, they approach each other and fuse, their endothelial linings becoming continuous to form a network of irregular capillaries. Capillary plexuses also develop within the embryo itself, and eventually – by about the 21st day – a capillary network extends from the embryo to the trophoblast, via the body stalk, and ventrally into the wall of the yolk sac. The next important stage in the development of the vascular system consists of the formation of a functional heart and the selective enlargement of certain pathways within the network to form major blood vessels.

Development of the heart

The heart develops from initially isolated groups of mesodermal cells scattered in a U-shaped pattern around the future head end of the embryo. The cells are placed quite peripherally, lying beyond the buccopharyngeal membrane and the margins of the neural plate. They occur only in the visceral mesoderm close to the endoderm, and dorsal to them is part of the intraembryonic coelom that will subsequently form the pericardial cavity. The future heart cells form a U-shaped plexus of endothelial vessels that communicates with the other capillaries developing in the embryo. The vessels of the lateral limbs of the heart plexus merge to form relatively substantial *right* and *left endocardial tubes*. (*See Figure 31a*.)

Following closure of the brain region of the neural tube, the head end of the embryo mushrooms out into the amniotic cavity and the peripheral margins of the embryonic disc become tucked ventrally beneath the developing body. The heart is thus moved to a new position ventral to both the neural tube and the foregut, and it now lies caudal to the buccopharyngeal membrane. Another change associated with this repositioning is the gradual approximation of the right and left endocardial tubes towards each other in the midline, until eventually they fuse to form a single median *endocardial tube* (*see Figure 31b*). The endocardial tube soon becomes invested with a thick layer of mesodermal cells, which will later differentiate into the muscle cells of the heart and its outer covering. Thus, by the end of the 3rd week of development the heart consists of a single thick-walled tube lying beneath the pharyngeal region of the foregut. Part of the tube lies within a bar of mesoderm crossing the embryo transversely between the future thoracic and abdominal regions (the septum transversum, which will later take part in the formation of the diaphragm), while the remainder of the heart tube lies in the primitive pericardial cavity. The cranial end of the

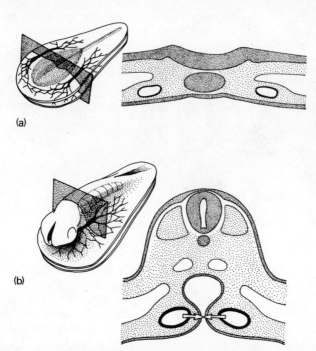

Figure 31. Early stages in the development of the heart. (a) Endocardial tubes and capillary plexus in the visceral mesoderm. (b) Approximation of the endocardial tubes as neurulation nears completion.

heart is linked with the network of vessels that will develop into the arterial system, and the caudal end is continuous with the future venous network.

The heart of the human embryo is thought to begin beating on approximately the 23rd day after conception. In more readily studied embryos, it has been observed that the first heart beats are irregular and weak and that the embryonic blood simply moves in and out of both ends of the heart with each beat. Gradually the activity of the heart tube becomes more regular and forceful, and blood begins to flow in one direction through it, entering at the caudal end and leaving at the cranial end. Thus, circulation of the blood begins, and major pathways are established in the intra- and extraembryonic capillary networks. By the 4th week blood leaving the heart enters vessels that arch around the pharynx and join two parallel longitudinal arteries called the *dorsal aortae*. In addition to intraembryonic branches, branches of the aortae extend to the yolk sac — *vitelline arteries* — and via the body stalk to the trophoblast — *umbilical arteries*. The blood carried away from the heart by this primitive arterial system is returned to the heart via the *umbilical veins*, the *vitelline veins*, and the *cardinal veins* that drain the embryonic body. (*See Figure 32.*)

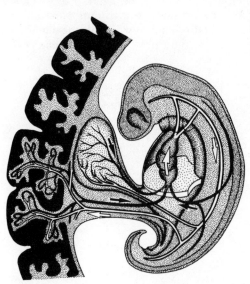

Figure 32. The pattern of blood circulation in a 4-week embryo. The black arrows indicate the return of blood to the heart through venous channels from the placenta, yolk sac, and embryonic body, while the white arrows indicate the flow of blood through the heart and arterial routes.

The heart tube, which now lies in the region of somites seventeen to twenty, develops dilatations that are forerunners of the later subdivisions of the heart. Starting at the caudal end and moving cranially, these are the *sinus venosus*, the *atrium*, the *ventricle*, and the *bulbus cordis* (*see Figure 33a*). Initially the sinus venosus and the atrium lie just outside the pericardial cavity in the septum transversum, but as a result of the events that follow they eventually come to lie within the cavity.

The bulbus cordis and the ventricle begin to elongate rapidly, and, since both ends of the heart tube are effectively anchored by substantial mesodermal structures, the relatively free region of the tube within the pericardial cavity bends. As growth continues, the heart appears first as a simple loop with its convexity

A Description of Normal Development

directed ventrally and to the right, and then as an S-shaped structure with the atrium (now within the pericardial cavity) lying dorsal to the ventricle (*see Figure 33b*).

The atrium is enlarging rapidly at this stage, and it becomes moulded around the bulbus cordis, showing the first signs of its conversion into two separate chambers. This moulding process is accompanied by internal changes — the formation of septa — that continue the subdivision. The septation of the atrium and ventricle is shown in *Figure 34*.

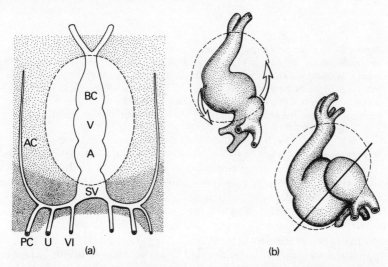

Figure 33. Subdivisions of the primitive heart tube, and formation of the cardiac loop. (a) The primitive heart tube and its subdivisions: SV = sinus venosus, A = atrium, V = ventricle, and BC = bulbus cordis. The major veins draining into the heart are also shown: AC = anterior cardinal, PC = posterior cardinal, U = umbilical, and VI = vitelline. Heavy stippling indicates the position of the septum transversum. (b) Formation of the cardiac loop. The line passing through the atrium and ventricle in the second drawing *(lower right)* gives the plane of section used in Figure 34.

Septation of the atrium

A septum develops from the internal surface of the atrium and extends across the central cavity. This is called the *septum primum*. Meanwhile, cellular proliferation in the walls of the relatively narrow canal joining the atrium and the ventricle produces two elevations, positioned diametrically opposite each other, which eventually meet and fuse in the midline. These *endocardial cushions* thus separate the original canal into two channels, and the growing septum primum subsequently comes into contact with them. However, before fusion occurs between the septum primum and the endocardial cushions, the central region of the septum thins and becomes degenerate with the result that a hole appears. This ensures that mixing of blood between one side of the atrium and the other can continue, the significance of which will be explained later. Another septum, the *septum secundum*, begins to develop from the atrial wall just to the right of the

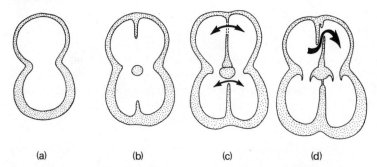

Figure 34. Septation of the atrium and ventricle. The atrium is above, the ventricle below. (a) Before septation. (b) Septum primum, endocardial cushions, interventricular septum. (c) Perforation of the septum primum. There is mixing of blood between the left and right atria and the left and right ventricles. (d) Downgrowth of the septum secundum to complete the foramen ovale and formation of the membranous part of the interventricular septum.

original septum. It is thicker than the first and grows parallel to it until it overlaps the perforation in the septum primum. However, growth of the second septum ceases before the lower edge contacts the endocardial cushions, and its major function is to act as a flap-like valve over the perforation in the septum primum. The two septa form a partition within the atrium and divide it into two *atria*. The perforation, with its valve, is called the *foramen ovale*.

The sinus venosus consists of a left and a right horn, in effect the remnants of the original pair of endocardial tubes. During development of the atria, unequal growth of the sinus horns results in their discharging the venous blood that they carry into the right atrium only. Finally, the right horn of the sinus is incorporated into the wall of the right atrium and the left horn remains as a wide venous structure draining the substance of the heart, the *coronary sinus*.

Septation of the ventricle

During the period of septation of the original atrium, a septum also develops in the ventricle and grows towards the endocardial cushions. This *interventricular septum* is thick and muscular and ceases growing just short of the endocardial cushions, leaving an interventricular foramen. Thus two ventricles are formed from the primitive ventricle. At the same time, a *spiral septum* develops in the distal part of the bulbus cordis from two opposite elevations and separates the outflow from the heart into two channels curved around each other, as shown in *Figure 35*. These channels later become the definitive *aortic trunk* and *pulmonary trunk*. The septum which subdivides the remainder of the bulbus cordis fuses with the interventricular septum, and thus the blood from the right ventricle is channelled into the pulmonary trunk and blood from the left ventricle is channelled into the aortic trunk. The defect in the interventricular septum is finally closed by the proliferation of tissues in the endocardial cushions and bulbar septum, so that total separation of the left and right ventricles occurs before birth.

Therefore development of the heart follows a complex sequence. The two primitive endothelial tubes are carried to a ventral position where they fuse to form a midline tube. Dilations subdivide the tube into a linear progression of chambers,

A Description of Normal Development

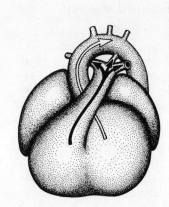

Figure 35. Formation of the spiral septum in the distal part of the bulbus cordis. This subdivides the outflow from the heart into an aortic and a pulmonary component. Blood leaves the right ventricle via the pulmonary trunk (black arrows) and the left ventricle via the aortic trunk (white arrows).

and at this stage the muscle fibres in the heart wall begin to function. The tube becomes bent into an S shape as a result of rapid growth in a confined space, and the major chambers become split into right and left components by the formation of internal septa. The heart has by this time achieved its definitive form, with venous blood returning to the right atrium and passing into the right ventricle from which it is pumped into the pulmonary trunk, while blood passes from the left atrium to the left ventricle and thence to the aortic trunk. However, there is still communication between the atria via the foramen ovale, and most of the blood entering the right atrium passes directly into the left atrium through the foramen. Before birth, the return of blood from the lungs to the left atrium is very slight as a result of the nonfunctional nature of the lungs. Once the two-sided heart has been established, a number of detailed changes occur — e.g. development of the atrioventricular, aortic, and pulmonary valves and elaboration of the conducting system of the heart — but the basic plan remains unchanged.

The most commonly encountered congenital abnormalities of the heart relate to the process of septation of the original heart tube. It appears that abnormalities occurring earlier than this would result in death of the embryo due to circulatory insufficiency, except in very rare cases in which one embryo becomes dependent on a normal twin for its blood supply and can survive even if it does not itself possess a functioning heart. Faults in septation — 'holes in the heart' — may range from a minor defect in the region of the foramen ovale to a wide communication between the atria or between the ventricles. An interesting pattern of malformations stems from an unequal subdivision of the bulbus cordis. The lumen of the pulmonary trunk is unduly small, and the ventricular septum fails to close. The enlarged aorta straddles the incomplete septum and receives blood from both ventricles. The blood pressure in the right ventricle becomes higher than usual because it is now competing with the powerful left ventricle, and eventually the right ventricle becomes hypertrophied. When these four abnormalities occur in combination — pulmonary stenosis, ventricular septal defect, overriding aorta, and hypertrophied right ventricle — they are referred to as the *Tetralogy of Fallot*. The defects cause a serious mixing of blood from the left side to the right, and so the body receives only partially oxygenated blood with signs of cyanosis becoming apparent soon after birth.

Development of the major blood vessels

In the 4-week embryo, blood leaving the median heart tube is carried around the developing pharynx within arching bars of mesoderm to join the dorsal aortae. The mesodermal bars are called the *pharyngeal arches,* and they give the neck region of the embryo a corrugated appearance at this stage. There appear to be six pairs of pharyngeal arches, but the sixth arch is poorly developed and difficult to delineate. A blood vessel develops in the core of each arch, and because of their shape and connections the vessels are called *aortic arches.* Thus six pairs of aortic arches are formed. However, they appear sequentially from the cranial to the caudal end of the pharynx, and at no time are they all present together. By the time the fifth and sixth arches are appearing the first arch is already so modified that its original form is lost. The changes that occur in this simple and symmetrical system of aortic arches to produce the more familiar pattern of vessels in the adult are shown in *Figure 36*. Note that the right and left *pulmonary arteries* arise from the proximal parts of the sixth arches. The remainder of the sixth arch on the right disappears, but on the left the remainder of the sixth arch becomes an important pathway, or shunt, linking the pulmonary circulation with the aortic flow. This linking vessel is called the *ductus arteriosus,* and before birth it carries

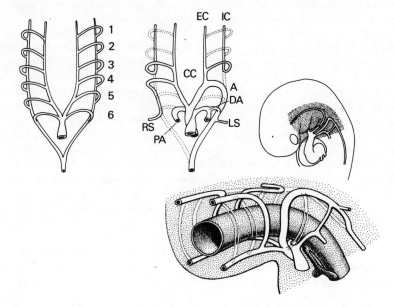

Figure 36. Development of the major arteries. *Upper left:* The primitive arrangement with six pairs of aortic arches feeding into paired dorsal aortae. *Upper centre:* The definitive plan, built mainly from arches 3, 4, and 6. The main arteries are shown: A = aorta, CC = common carotids, EC = external carotids, IC = internal carotids, PA = pulmonary, RS = right subclavian, and LS = left subclavian. The ductus arteriosus (DA) shunts blood from the left pulmonary artery to the aorta. *Upper right:* A key diagram to explain the orientation of the drawing below. *Lower right:* The relationship between the major arteries and the embryonic pharynx.

blood from the pulmonary trunk to the aorta since the nonfunctional lungs do not require the total output of the right ventricle before birth.

The dorsal aortae first appear as a pair of vessels running parallel to each other throughout their length, but as development progresses they fuse in the midline from the cardiac level caudally to form a single median vessel. The definitive descending thoracic aorta and abdominal aorta are derived from this vessel, and branches supply blood to the developing gut, the body wall, the limb buds, and the placenta.

Initially the venous return to the heart consists of a pair of vitelline veins from the yolk sac, anterior and posterior cardinal veins draining the embryonic body, and a pair of umbilical veins returning oxygenated enriched blood from the placenta. All these vessels converge on the right and left horns of the sinus venosus, which lie embedded in the septum transversum just caudal to the heart. The changes that now occur in this simple venous system are less clear-cut than the changes that take place in the arterial system, and there are commonly wide variations from the general plan. Usually, however, these variations do not greatly affect the efficiency of the system and are not therefore classed as malformations. This wider variation in the development of the veins compared with the development of the arteries may be a consequence of the relatively low pressure of blood in the venous system compared with that in the arteries; compare, for example, the meandering path taken by rivers crossing a coastal plain with the straighter cuts made by rapidly flowing water on a hillside. Development of the major veins is summarized in *Figure 37*. The right umbilical vein carries less and less blood as development proceeds, and eventually it atrophies completely leaving the left umbilical vein with the task of conveying all the blood from the placenta to the embryo. The passage of this enriched blood through the septum transversum is at first hindered by the network of sinusoids in the developing liver, but soon a bypass is established between the left umbilical vein and the right vitelline vein as it enters the sinus venosus. This bypass – the *ductus venosus* – acts as an important shunt before birth, but after birth, when the placental circulation ceases, it becomes obliterated and remains only as a ligament.

Foetal circulation

It is clear that the foetal circulation of blood differs significantly from circulation after birth. The foetal system incorporates a placenta and three shunts or bypasses: the ductus venosus, the foramen ovale, and the ductus arteriosus (*see Figure 38*). It is worth noting the foetal blood flow at this point, so that the changes that occur at birth (*see* p. 101) may be understood.

Blood returning to the foetus via the umbilical vein is about 80 per cent saturated with oxygen. It passes through the ductus venosus and joins with blood carried by the inferior vena cava, a vessel derived in part from the right vitelline vein and carrying deoxygenated blood from the liver, gut, and caudal regions of the body. Thus the blood entering the sinus venosus is now only 67 per cent saturated. Because of the anatomical arrangement of the right atrium, most of this stream of blood is directed towards and through the foramen ovale into the left atrium. On the other hand, deoxygenated blood returning from the head, neck, and upper limbs via the superior vena cava (derived mainly from the right

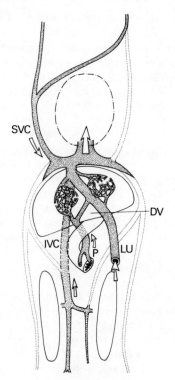

Figure 37. The developing venous system – a schematic diagram. Originally symmetrical, the return of blood to the heart eventually becomes channelled mainly into the right sinus horn. The major veins are shown: LU = left umbilical, which leads into the ductus venosus (DV), P = portal, IVC = inferior vena cava, and SVC = superior vena cava. The two oval outlines at the caudal end of the embryo represent the mesonephros.

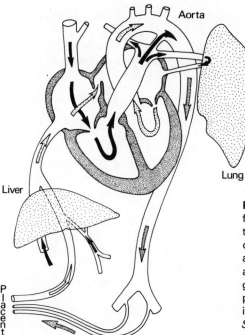

Figure 38. Circulation of blood in the foetus. The arrows indicate the direction of blood flow and the degree of oxygenation of the blood. The black arrows represent deoxygenated blood, and the white arrow represents oxygenated blood returning from the placenta. Stippled arrows indicate intermediate degrees of oxygenation. *See* text for a description of the foetal circulation.

anterior cardinal vein) is channelled predominantly into the right ventricle and thence into the pulmonary trunk. The oxygenated blood in the left atrium passes to the left ventricle and is pumped out into the aorta to be distributed to the foetal body and the umbilical arteries. The oxygen saturation of the blood as it leaves the heart is about 62 per cent, and the major arteries developed from the proximal part of the aorta carry blood with this composition to the developing head and upper limbs. The remainder passes down the descending thoracic aorta where it is joined by blood from the pulmonary circulation as it is shunted through the ductus arteriosus. Thus once again the oxygenated blood is mixed with deoxygenated blood, so that the blood distributed to the caudal half of the body is poorly oxygenated compared with that going to the head and upper limbs. This may explain the difference in growth rates between the upper and lower halves of the body. The umbilical arteries arise from the caudal end of the abdominal aorta where it divides to form the common iliac arteries. They carry the blood not distributed within the foetus through the umbilical cord to the placenta where waste products can be discharged into the maternal blood and nutrients and oxygen can be collected.

The Face

The development of the face from the 4th week to the 20th week is illustrated in *Figures 39, 40, and 41*.

The face is built up around the primitive mouth by the growth and fusion of a number of mesodermal aggregates, which first appear early in the 4th week. These are the *frontonasal process*, on the ventral surface of the brain, and the paired *maxillary and mandibular processes*, which develop from the first pharyngeal arches. The maxillary process on each side grows medially from the cranial aspect of the first arch and extends between the developing eye and mouth region, while on the other side of the mouth the mandibular processes also grow medially and fuse in the midline to form the precursor of the lower jaw.

Two *olfactory pits* develop in the frontonasal process as it advances caudally towards the mouth. The edges of these pits become elevated and for a time have the appearance of comma-shaped ridges. Rather arbitrarily the elevations are subdivided for descriptive purposes into medial and lateral components. The maxillary processes eventually fuse with the components of the frontonasal process to complete the formation of the upper jaw. The mouth is initially very wide, but fusion of the lips occurs at their lateral angles and the mouth assumes a less grotesque and more recognizable shape. The eyes migrate medially from the sides of the head until they are forward facing, and the external ears move cranially from their original position on the sides of the neck, organized around the cleft between the first and second arches, to the more familiar position almost level with the eyes.

The olfactory pits deepen until they open into the cavity of the primitive mouth. Later, separate nasal and mouth cavities are established by the development of the palate. On each side a plate of tissue developed from the maxillary process grows medially into the primitive oral chamber. This is called the *palatal process*. Normally the two processes meet and fuse in the midline to form the hard palate. Initially, they are prevented from doing this by the presence of the developing tongue between them, but during the 8th week the embryo begins to preform mouth opening reflexes and these movements remove the tongue from between the palatal processes, allowing them to fuse. Meanwhile, a midline *nasal septum* extending between the olfactory pits grows caudally to meet the palatal shelves and fuses with the newly formed palate. Each nasal cavity now communicates with the exterior through the nostril and with the pharynx through an opening called the *choana*. The *soft palate* is formed by the growth of tissues from the posterior edge of the hard palate.

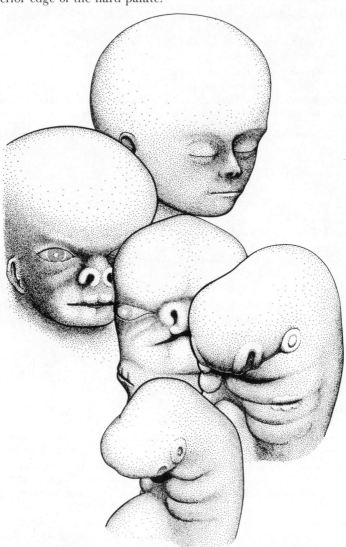

Figure 39. Development of the face. The stages shown are: 4 weeks *(bottom)*, 5 weeks, 7 weeks, 10 weeks, and 14 weeks *(top)*.

A Description of Normal Development 55

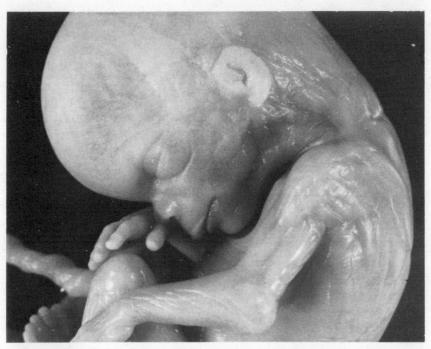

Figure 40. The face and hands at 12 weeks.

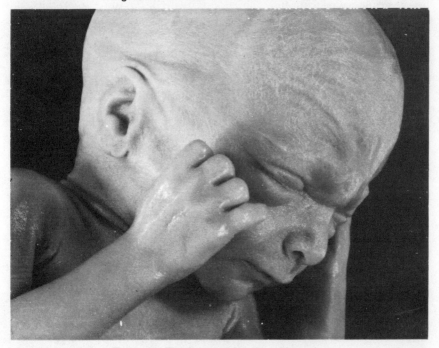

Figure 41. The face and hands at 20 weeks.

A number of facial abnormalities can be attributed to a failure of the several mesodermal aggregates to fuse. The most common of these — *cleft lip* and *cleft palate* — may occur separately or in combination. The incidence of babies having one or both conditions is approximately 1:800. Unilateral cleft lip, usually involving the upper lip, is caused by a failure of the maxillary process to fuse with the median nasal process on one side. In a rare condition called *oblique facial cleft* the deficiency extends as far as the eye, and this is caused by a complete lack of fusion between the maxillary process and the lateral and median nasal processes. Cleft lip is usually treated by surgery within the first two months after birth. Cleft palate is commonly associated with cleft lip. It is caused by a failure of the palatal processes to fuse with each other. A baby with severe cleft palate has difficulty in feeding and has to be nursed carefully until plastic surgery can be performed, generally when the infant is 1 — 2 years old.

The Eye

The receptive surface of the eye, the neural retina, is derived from the wall of the primitive brain. Thus the optic nerve, a bundle of fibres carrying the partially processed visual information from the retina to the brain, is in reality a *tract of brain tissue* rather than a nerve — strictly, a nerve is a collection of nerve fibres outside the central nervous system. The thin pigmented layer lying external to the retina is also derived from the original outgrowth of the brain, but other components of the eye (e.g. the lens and the outer sclera with its transparent corneal window) are formed from other sources. This mixed origin of the various parts of the eye necessitates a sequence of integrative or *inductive* processes during development, and this, together with the early appearance of the eye in the embryo and the ease of experimental manipulation, has made the eye a popular model for the study of developmental mechanisms.

The formation of the optic cup and lens vesicle is illustrated in *Figure 42*. Outgrowth of the brain wall begins very early. In the middle of the 3rd week shallow, ventrally directed bulges appear in the partially closed neural plate where it widens to form the future forebrain region. After closure of the neural plate, the bulges appear on the outer aspect of the forebrain as a pair of rounded protuberances, one on each side. These *optic vesicles* begin to enlarge steadily, and by the end of the 4th week the tip of each is in contact with the overlying ectoderm. It is thought that a slight positive pressure of fluid within the neural tube facilitates this expansion. The presence of the optic vesicle next to the ectoderm with only a narrow cell-free interlayer between them in some way triggers a sequence of changes that result in the formation of a lens vesicle. The changes resemble those seen during formation of the neural plate and neural tube. First, a local thickening of the ectoderm is produced by the elongation or 'palisading' of the ectodermal cells. This is achieved by the production of bundles of oriented microtubules, which push the nucleus towards the deep pole of the cell and cause elongation of the superficial region. The thickened disc of ectoderm formed in this way is called the *lens placode*. This response of the ectoderm to the optic vesicle is an example of an induced change; the optic vesicle is said to *induce* formation of the lens placode. The interrelationship has been investigated in numerous experiments, and it is now clear that there is some kind of chemical communication between the two layers that initiates the subsequent steps in the formation of the eye. It has been shown that if apposition between the optic vesicle and ectoderm does not occur, or if it is experimentally prevented from occurring, a lens does not form and the eye fails to develop.

The lens placode begins to invaginate soon after it is formed, partly as a consequence of nuclear migration and perhaps aided by contraction of apical microfilaments. It becomes first a saucer-shaped depression in the ectoderm and then a spherical hollow inpushing. The tip of the optic vesicle is affected by this invagination, becoming flattened and then pushed medially. It is as if the lens placode is invaginating with sufficient force to indent the optic vesicle, like a clenched fist being pushed into a soft balloon. Thus the optic vesicle is fashioned into a cup-shaped structure with a double wall, and the original lumen becomes reduced to a narrow cleft separating the two layers. Ventrally, the rim of the optic cup is deeply indented by the *choroid fissure*, a groove that provides a point of entry into the cup for the developing retinal blood vessels (*see Figure 42*). During the 7th week the lips of the groove usually close over the vessels to remedy this deficiency, but in rare cases this fails to occur and the cleft persists as a gap in the iris and/or retina.

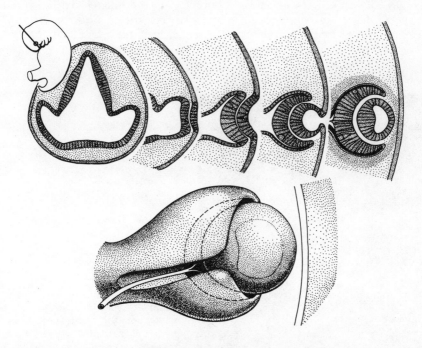

Figure 42. Development of the eye. *Top:* Formation of the optic cup and lens vesicle, shown in a series of transverse sections. *Bottom:* The arrangement of the choroid fissure.

58 Before Birth

The inner layer of the optic cup thickens as it invaginates and begins to differentiate into the *neural retina* with its four distinct layers of neural cells. However, the outer layer of the cup becomes thinned to a membrane composed of a single layer of cuboidal cells in which pigment granules soon develop, converting the posterior wall of the optic cup into the *pigment layer* of the eye.

During the 5th week the spherical lens rudiment finally separates inwards from the ectoderm and begins its next phase of development as a hollow *lens vesicle* situated in the mouth of the optic cup. The cells of its posterior wall elongate and form long *primary lens fibres*, which eventually fill the lumen of the vesicle. In this way the 'nucleus' of the lens is formed, and from the 7th week onwards growth of the lens is achieved by the addition of *secondary lens fibres* around the nucleus. It is believed that this process continues throughout life, although at a steadily diminishing rate.

The double-layered lip of the optic cup contributes to the formation of the iris and the ciliary body, structures that are completed by the addition of mesodermal cells that later differentiate into muscle fibres and connective tissue. The equator of the developing lens becomes linked with the ciliary body by a network of elastic fibres — the *suspensory ligament* — so that contraction of the ciliary muscle results in changes in tension in the ligament and alters the curvature of the lens.

The eye primordium is completely surrounded by mesodermal cells that segregate into two layers: an inner layer resembling the pia mater of the central nervous system, and an outer layer resembling the dura mater. The inner layer becomes a highly vascularized and pigmented layer called the *choroid*, while the outer layer develops into the tough outer shell of the eye called the *sclera*. Anteriorly, the sclera is modified to form the transparent cornea.

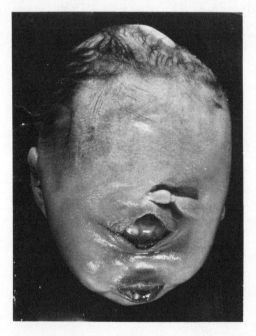

Figure 43. Cyclopia. In this rare abnormality, both the orbits are joined and there is a single median eye. Two corneas are apparent in this example, and there is a tubular proboscis over the eye indicating grossly abnormal development of the nose and nasal cavities.

Several abnormalities of the eye can occur. Persistence of the choroid fissure has been mentioned already, but in addition it is possible for the eyes to be abnormally small (microphthalmia), completely absent (anophthalmia), or fused to form a single median eye (cyclopia). A case of cyclopia is shown in *Figure 43*. The lens of the eye may become opaque before birth − *congenital cataract* − either as a result of maternal infection with rubella during the 2nd month of pregnancy or due to genetic factors.

The Ear

The human ear can be subdivided into three parts: the *external ear*, the *middle ear*, and the *internal ear*. The external ear is basically a sound-collecting funnel and passageway leading into the eardrum. The sound energy picked up by the eardrum is transmitted through three tiny bones in the middle ear chamber to the receptor mechanism of the internal ear, an elaborate system of fluid-filled chambers and canals called the *membranous labyrinth*. The sound-processing portion of the membranous labyrinth consists of a spiral-shaped *cochlea*, but closely associated with this is the *vestibular apparatus*, which is concerned with balance.

The development of the ear is illustrated in *Figure 44*. The precursor of the membranous labyrinth (the first part of the ear to arise) becomes visible during the 3rd week of development as a thickening of the surface ectoderm on each side of the folding neural plate. This thickening, the *auditory placode*, is thought to arise in response to induction by the future hindbrain region. In the 4th week the auditory placode invaginates in a way that resembles invagination of the lens placode and forms a closed *auditory vesicle* below the ectoderm.

Soon after its formation the auditory vesicle enlarges and becomes elongated dorsoventrally. The dorsal part expands, and conspicuous flanges appear: primordia of the *semicircular canals*. As the flanges push out from the main vesicle the walls of their central portions come into apposition and then undergo resorption, so that the original semilunate flanges become converted into loop-like ducts. There are three ducts formed in this way, each orientated at right angles to the other two. During this activity, the remainder of the vestibular portion of the auditory vesicle becomes subdivided into a dorsal *utricular* portion and a ventral *saccular* portion. The semicircular canals are related to the utriculus, and near one of their two openings into the utriculus each canal forms a local enlargement called the *ampulla*. Specialized receptors develop within the ampulla. They consist of hair-like processes extending into the lumen of the canal and are stimulated by fluid currents produced within the semicircular canals when the head moves. In this way the brain becomes aware of changes in position and acceleration of the head. In addition there are receptors within the sacculus and utriculus that give information about the *static* orientation of the body.

The sound-perceiving mechanism of the ear develops in the *cochlear* portion of the membranous labyrinth. The cochlea elongates rapidly during the 6 − 8th weeks and develops into a spiral with two and a half turns. A *spiral ganglion* of nerve cells develops alongside the cochlea and supplies fibres to the tonoreceptive *organ of Corti* within.

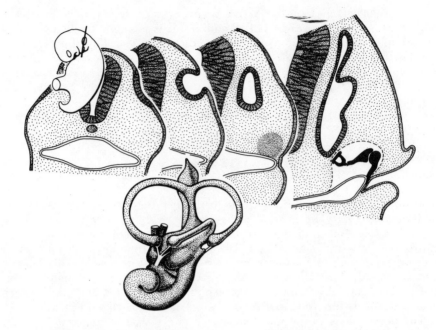

Figure 44. Development of the ear. *Top:* A series of transverse sections illustrating the formation of the auditory vesicle from the surface ectoderm and its relationship with the hindbrain, the first pharyngeal pouch, and the developing ossicles. *Bottom:* An intermediate stage in the differentiation of the auditory vesicle. The semicircular canals and spiral cochlear duct are already apparent.

While the inner ear develops, the transmitting apparatus of the middle ear is also being established. The cavity of the middle ear, together with the slender canal that links it with the pharynx, is derived from the first pharyngeal pouch (*see* p. 64). The *tympanic membrane*, or eardrum, is formed in the region where the endoderm at the tip of the pouch lies close to the surface ectoderm of the first pharyngeal cleft. The dorsal end of the skeletal element of the first arch forms the *malleus* and *incus,* and that of the second arch forms the *stapes,* thus making the chain of three auditory ossicles that transmits sound energy from the eardrum to the inner ear. At first the ossicles lie above the cavity of the middle ear, but towards the end of the foetal period the loose connective tissue around them is resorbed. The tympanic cavity enlarges, and its endodermal lining becomes moulded around the ossicles, which now arch across the chamber. Full movability of the ossicles is only achieved several months after birth.

The pinna of the external ear is formed by the coalescence and modelling of several mesodermal aggregations clustered around the first pharyngeal cleft.

A Description of Normal Development 61

The Digestive System

Formation of the primitive gut

The lining of the digestive tract is formed almost entirely from the endodermal layer of the embryo. During the 3rd week of development this layer forms the roof of the yolk sac cavity and is continuous with the lining of the yolk sac (*see Figure 45*, which shows the primitive digestive system at the 4th week). As the neurulating embryo mushrooms out into the amniotic cavity, the endodermal layer becomes gathered in ventrally to establish the *foregut, midgut,* and *hindgut,* and parts of the wall of the yolk sac are incorporated into the primitive gut. At first the midgut has a wide communication with the remainder of the yolk sac, and this enables the arbitrary distinction between foregut, midgut and hindgut to be made, but as the developing gut becomes drawn more deeply into the embryonic body the connection becomes narrowed and persists as an insignificant duct within the umbilical cord.

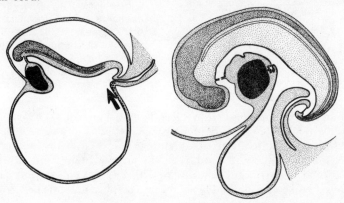

Figure 45. Initial stages in development of the digestive system. *Left:* A longitudinal section through an embryo in the middle of the 4th week. The black arrow indicates the direction of migration taken by the primordial germ cells. The large area of black represents the pericardial cavity. *Right:* The stage reached by the end of the 4th week. The connection between the primitive gut and the yolk sac has become narrowed as the head and tail ends of the embryo curl ventrally.

The position of the mouth is marked initially by a depression in the surface ectoderm. The floor of the depression is composed of a bilaminar plate of cells – ectoderm externally and endoderm internally – called the *buccopharyngeal membrane.* This membrane separates the lumen of the foregut from the amniotic cavity, but it breaks down during the 3rd week of development. At the tail end of the embryo a similar depression occurs, and once again the floor of the depression is formed of apposed ectoderm and endoderm. This *cloacal membrane* breaks down later to produce the anal and urogenital orifices. A diverticulum grows out from the hindgut into the body stalk to form the *allantois,* the proximal part of which becomes incorporated into the bladder.

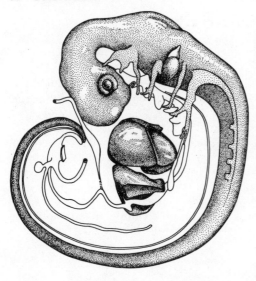

Figure 46. The digestive system in an embryo at the end of the 4th week. In this reconstruction, the central nervous system, heart, and liver are stippled, together with the rudiments of the eye and inner ear. Four cranial nerves are shown in light stipple. The pharyngeal region of the gut tube shows a number of outgrowths that will give rise to structures such as the thyroid and parathyroid glands, the thymus, and the respiratory system. The liver and nearby pancreas develop from endodermal buds at the caudal end of the foregut. In this embryo the midgut loop has lost its continuity with the vitelline duct. In the hindgut region there is still an open association with the allantois and the ducts of the urinary system that discharge into it. The cloacal membrane is intact at this stage. [Modified from Blechschmidt (1961). Compare with Figure 9, which gives an external view of the same embryo.]

Many other organs originate from the endodermal gut tube (*see Figure 46*). For example, diverticula from the tube give rise to the thyroid, parathyroids, thymus, liver, and pancreas and to the air-conducting passages of the respiratory system. In addition, the linings of the middle ear, the pharyngotympanic tube, parts of the urethra, and the prostate gland in the male are also endodermal. Thus the differentiation of the endodermal layer is very complex, and it is best considered region by region.

Development of the mouth and teeth

The cavity of the mouth is formed from the ectodermal depression described above together with the cranial end of the foregut. These two parts are at first separated by the buccopharyngeal membrane. Thus some of the structures in the mouth are derived from ectoderm while others are endodermal. The ectodermal structures include the coverings of the hard palate, the sides of the mouth, and the lips, together with the enamel of the teeth and the parotid salivary glands. Endodermal structures include the epithelium of the tongue, the soft palate, the floor of the mouth, and the sublingual and submandibular salivary glands. However, there is no visible sign of the line of demarcation in the adult.

The first sign of tooth development is seen during the 6th week. The ectoderm overlying the edge of each developing jaw becomes thickened to form the *dental lamina*. From the deep aspect of each lamina there arise ten *dental buds*, which grow into the adjacent mesoderm. The dental buds become hollow and cup-shaped, with a condensation of mesodermal cells — the *dental papilla* — within the hollow. Some of the mesodermal cells differentiate to form *odontoblasts*, which begin to lay down dentine inside the dental bud. The remaining cells in the dental papilla form the *pulp* of the tooth. Meanwhile the mesodermal cells surrounding the developing tooth become condensed to form a fibrous *dental sac*. The cells of the tooth bud close to the dentine now differentiate into *ameloblasts*, which build a layer of *enamel* on the outer surface of the dentine. When the enamel is completed, the cells of the original tooth bud become less conspicuous until they remain only as a delicate membrane covering the enamel, a membrane that is soon worn away when the erupted tooth is used for mastication.

The ectodermal cells forming the deep rim of the hollow dental bud grow even deeper around the dental papilla to form the *root sheath*, and dentine is laid down within it continuous with the dentine of the first-formed crown. Later the root sheath is replaced by a specialized form of bone called *cementum*, which is analogous to the enamel of the crown. When the root is complete, the fibroblast cells of the dental sac anchor the cementum to the bony socket in the jaw by the fibrous *periodontal ligament*. During the 3rd month the dental lamina produces a second generation of dental buds on the lingual side of the developing deciduous teeth. Thirty-two permanent dental buds are formed, compared with the twenty buds of the deciduous first generation, and they develop in a similar way to the deciduous teeth.

The deciduous teeth erupt into the oral cavity after birth, the lower central incisors usually appearing first at about six months and the remainder being visible by the end of the second year. The permanent teeth begin to erupt during the sixth year. The third molars are the last teeth to erupt, and they may not appear until the thirtieth year. It is not clear whether tooth eruption is the result of pushing forces or pulling forces. It has been suggested that growth of the root or increasing pressure within the pulp cavity may push the tooth out of its crypt. Others argue that the tooth is progressively pulled out of the socket by a shortening of the fibres within the periodontal ligament.

Development of the pharynx and adjacent structures

The pharynx undergoes a remarkable sequence of changes during development. In this region the endodermal lining of the foregut is separated from the ectoderm laterally and ventrally by arching bars of condensed mesoderm called the *pharyngeal arches*. There are thought to be six pharyngeal arches on each side, but the caudal arches are very small and indistinct, and therefore it is difficult to be certain about the exact number. Each of the major arches produces a swelling both on the surface of the embryo and within the foregut, with the result that a series of grooves can be seen between the arches both externally and internally. The ectodermal grooves are called *pharyngeal clefts* while the endodermal grooves are called *pharyngeal pouches*. This arrangement of arches and grooves gives the 4-week human embryo an appearance that is reminiscent of a primitive fish with gill arches and clefts (*see* Figure 9). However, this resemblance is only superficial, since gills are never present during human development and the grooves between adjacent pharyngeal arches do not penetrate the pharyngeal wall completely — they are only superficial indentations. Thus once again this similarity of the human embryo at a particular stage of development to simpler animal forms indicates a retention of useful developmental processes rather than a recapitulation of adult forms (*see* the introductory section, p. 3).

The mesodermal core of each arch differentiates into cartilage and bone, somatic muscle, and a primitive blood vessel. The arch also receives a nerve that grows into it from the hindbrain to innervate the musculature. The muscle tissue often migrates away from its arch of origin, but it retains the original cranial nerve supply. The *first pharyngeal arch* divides into two processes: the short *maxillary process,* which grows medially beneath the eye to form the upper jaw (maxilla) and upper parts of the face, and the longer *mandibular process,* which forms a cartilagenous bar in the region of the future lower jaw (mandible) (*see Figure 39*). The dorsal tip of the mandibular process separates into two fragments that will later form two of the three ear ossicles: the *malleus* and *incus.* This mode of development of the ear ossicles illustrates an interesting change in function: in some species of fish the equivalent bones form part of a complex articulation between the upper and lower jaws, and it is only in later forms that they have assumed an auditory function. During the 6th week of development the mandible begins to form by intramembranous ossification of the mesodermal core of the mandibular process, and much of the original cartilagenous bar disappears without taking part in its formation. Several muscles are developed from the mesoderm of the first arch, the most important being the muscles of mastication.

The *second pharyngeal arch* gives rise to a number of bony and ligamentous structures on each side: the third ossicle of the ear (the *stapes*), the needle-like *styloid process* projecting from the base of the skull, the upper part of the small *hyoid bone,* which lies just below the angle between the chin and the neck, and a ligament between the styloid process and the hyoid bone. The muscles of facial expression are among the muscles derived from the second arch.

The mesoderm of the *third pharyngeal arch* completes the hyoid bone and probably provides some of the muscles of the definitive pharynx.

The *fourth* and *fifth* (and sixth) *pharyngeal arches* contribute the cartilages of the larynx and trachea. The muscles derived from these arches become the muscles of the larynx, the soft palate, and the remaining muscles of the pharynx.

The changes that occur in the major artery of each pharyngeal arch are described on p. 50.

As the pharyngeal arches become modified and various tissues differentiate within the mesodermal core, changes occur concurrently in the clefts and pouches. The *first pharyngeal cleft* plays an important part in the development of the external ear by forming the external auditory meatus and by contributing the outer ectodermal epithelium of the eardrum. Six small elevations around the external opening of the meatus enlarge and fuse to form the *pinna* of the ear. The remaining pharyngeal clefts become buried as the second arch in particular grows rapidly and overshadows them. Eventually the clefts disappear.

The endodermal pouches give rise to many structures, some of which remain in the pharyngeal region while others migrate to different final positions (*see Figure 47*). A diverticulum from the *first pharyngeal pouch* extends laterally to meet the floor of the first pharyngeal cleft. As it does so, it envelops the developing ossicles of the ear and establishes the chamber of the *middle ear.* The endodermal lining provides an inner epithelial component to the eardrum. The neck of the recess narrows to become a tube linking the middle ear with the pharynx. This tube, the *pharyngotympanic tube,* is important postnatally since it enables the air pressure inside the middle ear cavity to be equalized with the external atmospheric pressure and so prevents damage to the delicate eardrum.

A Description of Normal Development 65

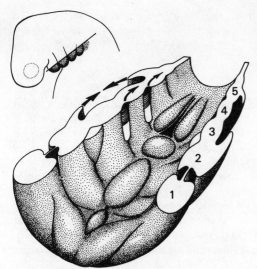

Figure 47. The embryonic pharynx. The floor and side walls of the pharynx are shown. The pharyngeal arches are numbered from 1 to 5 and are demarcated by clefts externally and pouches internally. The tongue is developing from a number of swellings in the floor of the pharynx. *Upper left:* Diagram indicating the plane of section. *See* text for explanation of the developmental processes that occur in this region.

Endodermal cells lining the *second pharyngeal pouch* proliferate and infiltrate the underlying mesoderm to form the palatine *tonsil:* a collection of lymphoid tissue that plays a defensive role against bacterial infection after birth.

The *thymus* and *inferior parathyroid glands* develop from the *third pharyngeal pouches*. The thymus arises from a pair of diverticula, one from each third pouch, which grow caudally through the neck to reach the ventral aspect of the definitive aorta. Here the two rudiments fuse to form a single midline structure and lose their connection with the pharyngeal pouches. The endodermal cells of the thymus are split into groups during the 3rd month by ingrowing septa of mesodermal cells, and increasing numbers of *lymphocytes* appear within the reticulum. By the time of birth the thymus is a relatively large organ, extending from the thyroid cartilage in the neck to the pericardial region in the chest. It continues to enlarge postnatally but at a slower rate than the body as a whole, so that it becomes restricted in extent to the upper part of the chest. It reaches its maximum size at the time of puberty and thereafter begins to regress. The thymus plays an essential role in establishing and potentiating the lymphatic system. If it is absent, the individual will be endangered by even the mildest of infections.

The endodermal cells of the future inferior parathyroids accompany the thymic diverticula during the first part of their migration, but on reaching the region of the thyroid gland the parathyroid tissue separates from the thymic tissue and remains closely applied to the posterior aspect of the thyroid gland.

Proliferation of the endodermal cells of the *fourth pharyngeal pouch* on each side marks the beginning of development of the *superior parathyroid gland*. The cells migrate caudally, losing contact with their pouch of origin, but they do not travel as far as the cells of the inferior parathyroid glands and come to rest in a more cranial position on the posterior surface of the thyroid gland.

There is disagreement about the existence or otherwise of a fifth pharyngeal pouch. Those who maintain that it exists suggest that some of the endodermal cells lining it migrate to the thyroid gland where they will later secrete the hormone *calcitonin*. The greater part of the thyroid gland is developed from a diverticulum that grows caudally from the ventral floor of the pharynx. To appreciate the position of this diverticulum it is first necessary to consider the development of the tongue.

Development of the tongue

During the 4th week of development a midline elevation appears in the ventral wall, or floor, of the pharynx. Soon after the formation of this *tuberculum impar*, a *lateral lingual swelling* develops on each side of it and grows medially (*see Figure 47*). The three elevations eventually fuse to form the body of the tongue. Meanwhile, a second median swelling called the *copula* develops in the floor of the pharynx just caudal to the tuberculum impar. As the copula enlarges it extends cranially around the tuberculum and becomes V-shaped. Later the copula is largely superseded by a contribution from the rapidly growing third pharyngeal arch, from which most of the caudal one-third of the tongue is formed. The muscles of the tongue are derived from the occipital myotomes, which at first are closely related to the developing hindbrain but then migrate to the pharynx and enter the tongue.

Development of the thyroid gland

The thyroid diverticulum is formed between the tuberculum impar and the copula. It grows caudally through the underlying mesoderm, and its distal end becomes bilobed, eventually forming the body of the thyroid gland. By the 7th week the migrating tissue reaches its definitive position in relation to the larynx and trachea. The solid cord of cells linking the thyroid to the tongue usually fragments and disappears. Originally the thyroid gland consists of a solid mass of cells, but by the 3rd month the formation of characteristic thyroid follicles begins.

The complex changes that occur in the embryonic pharynx are not always completed successfully. Occasionally remnants of the pharyngeal clefts lie buried within the tissues of the neck, and these have a tendency to enlarge and become troublesome in early adult life, necessitating surgical removal. The migration of cells from the endodermal pouches may be incomplete or aberrant. For example, thyroid tissue sometimes becomes arrested at points along the normal path of descent between the tongue and the larynx, and there have been examples of abnormal migration to the thorax, the bronchi, or the oesophagus. The inferior parathyroid glands, because of their association with the migrating thymus, are occasionally drawn down into the chest.

Development of the oesophagus

This is developed from the narrow part of the foregut continuing caudally from the pharynx. It elongates rapidly as the heart and septum transversum shift tailwards during the 4th week, and muscle fibres develop in its wall. The muscle fibres in the upper one-third of the oesophagus are predominantly striated, and those in the lower one-third are mainly smooth, while in the middle third there is a blend of both types.

Development of the stomach

The stomach appears first as a fusiform dilatation of the foregut (*see Figure 46*). It is suspended in the intraembryonic coelom by the thin *dorsal* and *ventral mesenteries*. The dorsal margin of the primitive stomach grows more rapidly than the ventral margin, and it becomes concave ventrally. The convex dorsal margin is known as the *greater curvature,* and the concave ventral margin is called the *lesser curvature*. With further differential growth, and as a consequence of changes in the liver and midgut region, the stomach rotates to the right so that its original left side now faces ventrally and its right side faces dorsally. The ventral and dorsal mesenteries become modified by this rotation, the ventral mesentery spanning the short distance between the stomach and the liver to form the *lesser omentum* and the dorsal mesentery being pulled after the stomach to form a pouch known as the *greater omentum*. In the adult the greater omentum has the appearance of an apron hanging in front of (ventral to) the intestines.

A fairly common abnormality of the stomach is a constriction of its junction with the duodenum. This condition — *pyloric stenosis* — prevents the passage of food through the obstructed region, and the baby resorts to forcible vomiting. The muscle of the pyloric region is often overdeveloped, causing considerable narrowing of the central canal, and in severe cases it is necessary surgically to incise the hypertrophied muscle.

Development of the duodenum

The duodenum is formed from the caudal part of the foregut and the cranial part of the midgut. This portion of the gut grows rapidly and forms a C-shaped loop that is initially supported by dorsal and ventral mesenteries. As the stomach rotates to the right, the duodenum is also displaced and eventually comes to lie against the posterior abdominal wall to the right of the midline. The dorsal mesentery fuses with the lining of the intraembryonic coelom, so the duodenum is no longer mobile within the cavity. The liver and pancreas develop as endodermal outgrowths from the duodenum.

Development of the liver

This arises as a solid *hepatic bud* from the caudal end of the foregut and grows into the mesoderm of the septum transversum. Columns of endodermal cells grow from the bud, branching and anastomosing, and causing the vitelline and umbilical veins coursing through the septum to become broken up into sinusoids: thin-walled channels carrying blood through the septum. The mesodermal cells provide a capsule to the liver and proliferate within the substance of the organ, differentiating into blood cells.

A system of *bile capillaries* and *ducts* develops within the columns of endodermal liver parenchyma, and they converge on the *common hepatic duct,* which has developed from the narrow stalk of the hepatic bud. Bile secretion begins during the 5th month of development. The liver grows very rapidly and soon fills most of the primitive abdominal cavity. This is partly due to the haemopoetic (blood-forming) function of the liver from the end of the embryonic period to the 6th month of development, after which blood cells are generated predominantly in red bone marrow and the spleen.

The *gallbladder* develops as an outgrowth of cells from the hepatic bud. The end of the outgrowth expands to form the sac of the gallbladder while the stem of the outgrowth becomes canalized to form the *cystic duct,* which later will carry bile to and from the common hepatic duct.

Development of the pancreas

The pancreas is formed by the fusion of two buds: one initially ventral to the foregut, and the other dorsal to it. The *ventral pancreatic bud* originates, in common with the hepatic bud, close to the junction between the foregut and midgut. As the stomach and duodenum rotate, the ventral bud begins to approach the *dorsal pancreatic bud,* which is positioned slightly cranial to the ventral bud and extends into the dorsal mesentery. Eventually, fusion of the buds occurs and their internal systems of ducts become linked. Usually the main *pancreatic duct* is derived from the entire duct of the ventral bud and the distal part of the central duct in the dorsal bud, and the proximal part of the duct of the dorsal bud is obliterated. If the proximal part persists, it is referred to as the *accessory pancreatic duct.* The *pancreatic islets* arise as small buds from the developing duct system within the pancreas. Later they become isolated groups of cells and begin to secrete *insulin* in the 5th month.

Development of the remainder of the midgut

It is noted above that the distal half of the duodenum is developed from the first part of the midgut. The primitive midgut also gives rise to the remainder of the small intestine — the *jejunum* and *ileum* — and part of the large intestine. Bearing in mind the arbitrariness of these delineations into foregut, midgut, and hindgut, it is usual to state the large intestinal derivatives of the midgut as being the *caecum,* the *ascending colon,* and the first part of the *transverse colon.* After the narrowing of the communication between the midgut and the yolk sac, the midgut becomes a small-diameter tube that increases rapidly in length and forms a ventrally directed loop. This change necessitates an enlargement of the dorsal mesentery that joins the midgut to the dorsal body wall. In contrast to the stomach region, there is no ventral mesentery in the region of the midgut loop. The dorsal mesentery carries a number of arteries from the dorsal aorta to the gut and the yolk sac, the most significant one in this region being the *superior mesenteric artery.*

Rapid growth of the liver and the urinary system produces a shortage of space within the primitive abdominal cavity, so the relatively mobile midgut, which also is growing rapidly, is forced to *herniate* into the umbilical cord. This occurs during the 6th week, and as it is part of a normal developmental process it is referred to as 'physiological' herniation.

While the loop of the midgut is within the umbilical cord the proximal part becomes greatly coiled, but the distal part grows more slowly and remains uncoiled. A diverticulum forms just caudal to the apex of the loop and marks the site of the future *caecum.* A tiny finger-like outgrowth from this caecal diverticulum will become the *appendix.*

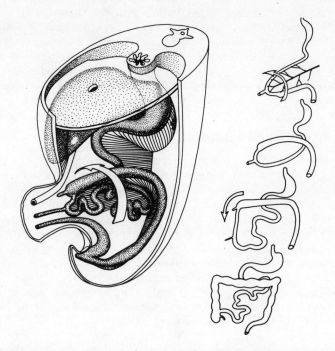

Figure 48. Rotation of the midgut. *Left:* Part of the abdominal wall has been removed in this diagram to show the contents of the abdomen. The cavity is limited cranially by the septum transversum, just caudal to which are the liver and stomach. The midgut loop has just started to return into the abdomen, and an intermediate stage of rotation is shown. *Right:* The complete sequence of positional changes is shown. [Modified from Langman (1975)]

Rotation of the midgut

During the 3rd month the abdominal cavity has grown sufficiently to reaccommodate the midgut, the return of which is illustrated in *Figure 48*. The herniated coils of the jejunum return first and come to lie high on the left side. Succeeding loops of the ileum then return, filling the left side of the abdominal cavity, and finally the caudal limb of the original midgut loop re-enters the abdomen. As the components of the midgut make this change, they rotate round an axis formed by the superior mesenteric artery. Viewing the embryo from a ventral aspect, this rotation occurs in an anticlockwise direction through an angle of 270 degrees. The caecum and appendix eventually come to lie ventral to the coils of jejunum and ileum that returned earlier, and positioned on the upper right side of the abdomen close to the right lobe of the liver as a result of the rotation that has occurred. In this way the physiological herniation of the midgut is corrected spontaneously and the derivatives of the original midgut loop are given their relative positions within the abdominal cavity. The caecum and appendix later move to a more caudal position until they come to rest in the lower right region of the abdomen.

Occasionally an abnormal rotation of the midgut occurs with the result that the intestines take on a different distribution within the abdomen. This may not cause any functional defects, but it may lead to mistaken diagnosis of the cause of abdominal pain. Another abnormality consists of the persistence of all or part of the duct linking the midgut to the yolk sac remnant. In its most severe form a canal from the ileum opens to the exterior through the umbilicus, but milder forms consist of cysts along the original path of the duct or a short diverticulum attached to the gut. Malformations of this type carry the risk of infection.

Development of the hindgut

The remainder of the digestive tract is formed from the primitive hindgut with a small contribution from the ectoderm in the anal region. Thus the *distal part of the transverse colon*, the *descending colon*, the *sigmoid colon*, the *rectum*, and the *upper part of the anal canal* are hindgut derivatives. At an early stage the caudal end of the gut is closed by the cloacal membrane formed from apposed layers of ectoderm and endoderm. Close to the cloacal membrane the hindgut sends off a diverticulum called the *allantois* into the umbilical cord, as shown in *Figure 45*. The slightly widened chamber formed where the allantois joins the hindgut is known as the *cloaca*. In the angle between their convergence there is a wedge-shaped bar of rapidly proliferating mesodermal cells — the *urorectal septum* — that grows caudally and fuses with the cloacal membrane, subdividing the cloaca into dorsal and ventral portions. The dorsal part becomes the *rectum* and the *upper part of the anal canal*, while the ventral part becomes the *urogenital sinus*. The lower part of the anal canal is formed from the ectodermal depression distal to the cloacal membrane. At a later stage the cloacal membrane perforates, and the lumen of the hindgut now opens to the amniotic cavity. In about 1:4 000 births a child is born with imperforate anus as a result of abnormal development of the urorectal septum or due to persistence of the cloacal membrane.

The Respiratory System

The respiratory system, the development of which is illustrated in *Figure 49*, develops embryonically as an outgrowth from the digestive tract — it has an endodermal origin. A small pocket-like diverticulum grows from the ventral wall of the pharynx during the 4th week and extends caudally into the visceral mesoderm on the ventral surface of the foregut. Distally this midline tube divides into left and right *lung buds*. The upper part of the respiratory diverticulum becomes modified to form the larynx. Its connection with the pharynx becomes narrowed, and laryngeal cartilages are formed from the skeletal elements of the fourth and sixth branchial arches. The lower part of the midline tube forms the trachea.

Each lung bud grows laterally and bulges into the intraembryonic coelom, the part that will later be modified to produce the fluid-lubricated pleural cavities separating the lungs from the thorax. The left lung bud divides into two lobes and the right into three, establishing the pattern of main bronchi and lobes that can be

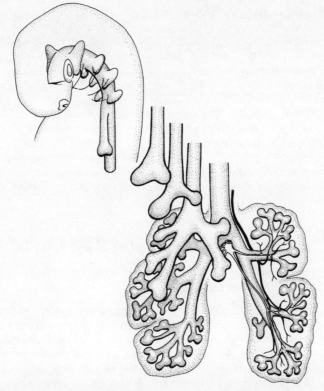

Figure 49. Development of the respiratory system. An endodermal outgrowth from the ventral aspect of the pharynx *(upper left)* gives rise to the lungs and airways by branching and associating with adjacent mesodermal cells.

observed in the fully developed lung. The growing tips of the bronchi then divide dichotomously, each branch again bifurcating to give two branches, until the tubular system within each lobe resembles the branching canopy of a tree. Branching continues throughout the foetal period and also for a time after birth until a full complement of terminal bronchioles and alveoli is formed. The process is completed by the time the child is about 8 years old. The mesodermal cells surrounding the endodermal bronchial tree give rise to the other components of lung tissue: muscle fibres, highly elastic connective tissue, cartilage, and blood vessels. Mesodermal cells also form the thin membranous linings of the pleural cavities.

Before the 7th month of development the foetal lungs are incapable of meeting the respiratory needs of a prematurely born baby. This is because the alveolar surface area is too limited and because the capillary loops connected with the pulmonary circulation are still being developed. Movements of the thoracic cage resembling future respiratory movements occur long before birth and have been observed in aborted foetuses only 3 months old. As a consequence of these rehearsal movements some amniotic fluid is drawn into the developing bronchial tree, and this is supplemented by the secretions of glands lining the passageways.

72 Before Birth

The foetal lungs therefore resemble waterlogged sponges. Before aeration of the lungs can be achieved at birth, the fluid must first be rapidly removed. Some of the fluid is forced out during birth as the baby passes down the birth canal, while the remainder is rapidly absorbed through the lining of the bronchial passageways and carried away by blood and lymphatic vessels. Thus aeration of the lungs at birth consists of a replacement of fluid by air; it does not consist simply of the inflation of compact collapsed organs as is sometimes mistakenly supposed.

The ease of fluid displacement and the efficiency of gaseous exchange in the lungs is increased by the presence of substances known collectively as *pulmonary surfactant*. Surfactant reduces the surface tension of the fluid film lining the alveoli. Synthesis begins during the 6th month of foetal development, but an adequate concentration is not immediately reached, and many cases of respiratory distress in premature babies have been attributed to an insufficiency of surfactant.

The Musculoskeletal System

Differentiation of the somites

During the period of formation and closure of the neural plate, the mesodermal layer becomes organized into three zones on each side of the midline: a column of somites, the intermediate mesoderm, and the lateral mesoderm composed of somatic and visceral layers. About forty-three pairs of somites are developed in a craniocaudal sequence, and at first they consist of solid aggregations of mesodermal cells. These somites can be seen in Figures 8, 9, and 10. The cells are temporarily reorganized and take on the form of a hollow wedge that has its base adjacent to the neural tube and is approximately square when viewed from above. Then three components can be discerned within each somite: two fairly compact layers dorsolaterally called the *dermatome* and the *myotome*, and a ventromedial portion called the *sclerotome* (see Figure 50). The sclerotomal cells soon begin to dissociate themselves from the somite, and they migrate to new positions

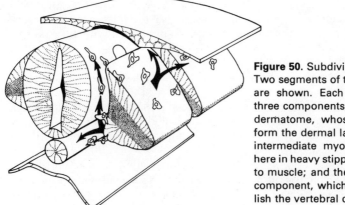

Figure 50. Subdivisions of the somite. Two segments of the embryonic trunk are shown. Each somite splits into three components: an upper part, the dermatome, whose cells migrate to form the dermal layer of the skin; the intermediate myotomal part (shown here in heavy stipple), which gives rise to muscle; and the lower sclerotomal component, which migrates to establish the vertebral column.

around the neural tube and the notochord. There they start to establish the vertebral column, or backbone, and it is from this role that the sclerotome derives its name (Greek *scleros* = hard). Later the cells of the dermatome (Greek *derma* = skin) also migrate from their original site and reform immediately beneath the surface ectoderm. Here, together with some of the somatic mesodermal cells of the body wall, they form the deep *dermal layer* of the skin, while the overlying ectoderm forms the *epidermis*.

Development of the muscles

Muscles of the body are of three main types: somatic or voluntary muscle, visceral or involuntary muscle, and cardiac muscle. The following description is limited to the development of somatic musculature.

The myotomes (Greek *mus, muos* = muscle or mouse) increase in thickness, and migration and reorganization of the constituent cells occur as in other parts of the original somite. After migration the myotomal cells begin to differentiate into muscle cells, becoming elongated and spindle-shaped. Microfibrils develop within the cytoplasm, and they become aligned longitudinally to form cross-striated bundles. Developing muscle cells have been observed to fuse with neighbouring cells with the result that large multinucleate muscle fibres are produced. Similar changes occur in the mesodermal cells of the lateral and anterior body walls and in the developing limb buds. The growth of muscle during the foetal period is due to three processes: the formation of additional myoblasts from mesodermal cells, the division of existing but incompletely differentiated myoblasts, and the hypertrophy of existing muscle fibres by an increase in the number of cytoplasmic myofibrils. After birth, muscle growth occurs only by hypertrophy of existing fibres. Groups of muscle fibres become enclosed in connective tissue sheaths derived from mesodermal cells, thus forming individual muscles that are attached directly to bones or via tendons, which also originate from condensations of mesodermal cells.

Many of the muscles of the head are developed from the mesoderm of the pharyngeal arches, while those of the tongue are thought to be derived from the three pairs of occipital myotomes placed very cranially in the somite rows. The extrinsic muscles of the eye, those that move the eyeball within the orbit, develop from three preotic myotomes, each supplied by its own cranial nerve.

The myotomes of the neck and trunk regions become divided into a smaller posterior part called the *epimere* and a larger anterior part called the *hypomere*. Each segmental spinal nerve also divides to accommodate this change, and the two branches are known as the *posterior primary ramus* and the *anterior primary ramus* respectively. The epimeres eventually develop into the extensor muscles of the back, while the hypomeres develop into muscles with a flexor action.

The muscles developed in the somatic mesoderm of the body wall form the anterior and lateral muscles of the neck and trunk. In the thorax the muscles generally retain their segmental arrangement and lie in the spaces between the ribs, but in the wall of the abdomen the mesoderm is uninterrupted by skeletal elements and large sheets of muscle are formed. In both the thorax and abdomen the muscular wall becomes organized into three layers of thin circumferential muscles, which can be clearly discerned, although with some local modifications, in the adult.

74 *Before Birth*

The muscles of the limbs are derived from the mesoderm of the limb buds, which begin their development in the 5th week after conception. The arm buds appear a little earlier than the leg buds and grow more rapidly during the embryonic period. This disproportionate growth is corrected later during the foetal period. The arm buds lie at the level of the lower six cervical and upper two thoracic segments, and the leg buds lie at the level of the lower four lumbar and upper three sacral segments. As the limb buds enlarge, the anterior primary rami of the spinal nerves situated close to their bases begin to grow into the limbs. Later the mesoderm in each limb separates into anterior and posterior portions, and the nerve trunks entering the limb buds also subdivide into anterior and posterior portions. Muscles are formed within the limbs by the migration of groups of mesodermal cells, and the limb buds themselves begin to move in relation to the vertebral column, so that the originally simple pattern of nerves entering the limb buds becomes reorganized to form complicated nerve *plexuses* near the base of each limb.

Development of the bony skeleton

After the sclerotomal cells have arrived at their definitive positions they gradually establish the bony skeleton. This is achieved in two ways, the method used depending on the type of bone that is to be built. One method is known as *intramembranous ossification* and the other is called *endochondral ossification*, both of which are illustrated in *Figure 51*.

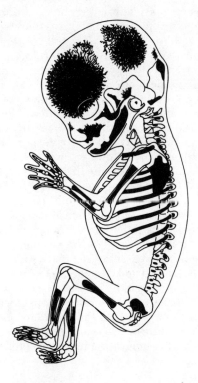

Figure 51. The pattern of ossification in a 4-month foetus. Developing bone was stained with alizarin and the soft tissues cleared in glycerin to display the extent of ossification. Note the difference in appearance between the intramembranous ossification in the bones of the cranial vault and the endochondral ossification in the limb bones.

Intramembranous ossification

Mesodermal cells form a dense aggregation at the site where this process is to occur, and they secrete collagenous fibres into the intercellular spaces, transforming the area into a fibrous membrane. Some of the cells differentiate into *osteoblasts*. These secrete a dense, also fibrous, material called *osteoid* and an enzyme called *alkaline phosphatase*, which causes the deposition of calcium within the fibrous matrix in the form of tiny *apatite* crystals. In this way the initially pliable osteoid is converted into a rigid bony material. Some of the osteoblasts become trapped by the expanding bony spicules and continue to live in tiny lacunae within the bone, now being referred to as *osteocytes*. The point at which these changes begin is known as an *ossification centre*.

A vascular layer, the *periosteum*, develops on each surface of the original membrane, and osteoblasts within the layer begin to deposit bone on the surface. The bone produced beneath the periosteum is more dense and compact than that being developed within the membrane. It is thus referred to as *compact bone*, in comparison with the spongy or *cancellus bone* being produced internally. Vascular tissue begins to fill the spaces within the cancellus bone and becomes *red bone marrow*, whose function is to generate new blood cells.

As development and growth of the foetus continue, it is important that the bony tissues keep pace and do not restrict the expansion of soft tissues. Thus it is necessary for existing bony structures to be continuously remodelled and reproportioned. Remodelling depends on two complementary processes: the resorption of existing bone by giant multinucleate cells called *osteoclasts*, and the laying down of new bone by osteoblasts.

Endochondral ossification

The long bones of the body develop in a different way from that described above, which is predominantly concerned with the formation of thin flat bones. After the condensation of mesodermal cells at the site of the future long bone, a *cartilagenous model* of the bone is formed. Then, as the foetal period begins, ossification of the model is initiated in the middle of the shaft. The cartilage cells enlarge, and calcium is deposited in the cartilagenous matrix. The trapped cells degenerate with the result that empty cavities are left within the ossification centre. A periosteum is established around the outer surface of the shaft, and this begins to form a sheath of compact bone in the region of the primary ossification centre. Periosteal blood vessels grow towards the zone of calcified cartilage, and with the aid of attendant osteoclast cells they ramify within the honeycombed region. Osteoblasts that have accompanied this vascular invasion lay down new bone within the cavities, so a core of cancellus bone is established within the shaft. As the shaft grows, remodelling processes produce a large central cavity that becomes filled with red marrow.

Ossification now advances away from the ossification centre towards the ends of the cartilagenous model. During this time the model itself continues to grow in proportion with general foetal growth. At birth a long bone has a shaft (*diaphysis*) that is almost wholly converted to bone, with the two ends of the original model (*epiphyses*) still cartilagenous. Later, one or more *secondary ossification centres* appear in each epiphysis, and the cartilagenous ends are converted into bone. Thus the cartilagenous component becomes reduced to a thin plate at each end of the shaft in the interface region between the steadily enlarging primary and secondary

76 *Before Birth*

ossification centres. These *epiphysial plates* are the sites at which postnatal growth occurs, and the growth in length of long bones ceases when the epiphysial plates are finally converted into bone. When this happens, usually by about the twentieth year, the epiphyses and diaphysis are fused to form a continuous bone. Growth in diameter is the result of the continued deposition of compact bone around the periphery of the bone beneath the periosteum, and as in the case of intramembranous ossification there are remodelling processes at work throughout the period of growth.

With these two types of ossification in mind, it is possible to consider how the skeleton is built up in different parts of the body.

Development of the skull

The skull consists of a protective bony case around the brain together with the bones of the face and jaws, and both types of ossification are in evidence. The different parts of the skull appear first as condensations of mesodermal cells, which later become converted into membrane, bone, or cartilage. Generally, the bones forming the base of the skull develop from cartilagenous models. Endochondral ossification begins early in the 3rd month with each bone arising from two or more centres of ossification. On the other hand, the large curving plates of bone that form the roof of the skull are intramembranous in origin. Most of the bones of the skull are ossified by the time of birth, but they are not fused edge to edge at this stage, thus allowing the bones to move relative to each other and even to overlap. This mobility is important during the process of birth because it enables the dimensions of the head to be reduced as it passes through the restricted birth canal. After birth a baby's head may look quite distorted, but the irregularities are soon corrected.

In regions where several bones of the skull approach each other there tend to be larger areas devoid of bone than in regions where only two bones lie side by side. The two most significant of these *fontanelles* occur on the top of the skull in the midline. They are the *anterior* and *posterior fontanelles*. By examining the fontanelles a doctor can determine the process of growth in adjacent bones, the degree of hydration of the baby (a depressed fontanelle indicates dehydration), and an impression of intracranial pressure (raised pressure causes bulging of the fontanelles). As ossification spreads from the surrounding bones the fontanelles shrink and eventually close, the smaller posterior fontanelle closing by the end of the first year and the anterior fontanelle closing by the end of the second year.

Development of the vertebral column

The migrating sclerotomal cells surround the notochord and neural tube. The cylinders of mesodermal cells produced still show signs of their origin from the segmental somites, but rather surprisingly the vertebrae are developed, not within the original segments, but *between* the segments, as shown in *Figure 52*. The caudal half of each sclerotomal unit fuses with the cephalic half of the succeeding sclerotome to form a vertebra. Each vertebra is therefore an intersegmental structure, and the segmental spinal nerves leave the vertebral column between adjacent vertebrae. This mode of origin of the vertebrae also allows the short segmentally arranged muscles of the back to span from one vertebra to

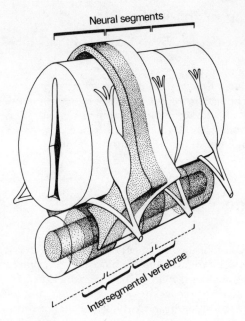

Figure 52. Development of intersegmental vertebrae from sclerotomal cells. Part of the neural tube is shown, together with the underlying notochord, which is ensheathed by sclerotomal cells. Three segmental spinal nerves arise from the neural tube, and the arch of a developing vertebra lies between two sets of nerve roots. Each vertebra is formed from the caudal part of one sclerotome (stippled) and the cranial part of the next (unstippled).

another and thus to produce movement at the joint in between. The notochord disappears completely in the region of each vertebral body, but in the intervertebral regions it contributes to the formation of the *intervertebral disc*, forming the centrally placed *nucleus pulposus*. The condensations of sclerotomal cells first produce cartilagenous models of the vertebrae complete with a *centrum*, a *neural arch*, and various *processes*. (In the thoracic region, long costal processes pass laterally between the myotomes to initiate development of the ribs.) At about the 9th week of development a number of primary ossification centres appear in each vertebra, generally two in the centrum and one for each half of the neural arch. Complete union of all the primary centres does not occur until several years after birth, and several secondary centres that are developed in the remaining cartilagenous areas of the vertebra (e.g. the tips of processes) do not fuse with the rest of the vertebra until about the twenty-fifth year.

The embryonic vertebral column is gently *curved*, appearing concave ventrally. During foetal development a slight convexity occurs at the lumbosacral junction. After birth other curvatures develop in response to changes in behavioural patterns. When the child begins to raise its head, a cervical curvature develops that is convex anteriorly (the term 'anterior' is used in descriptions of postnatal human anatomy in place of the more universal word 'ventral'). When the child begins to stand and walk, a lumbar curvature develops that also is convex anteriorly. These curvatures of the vertebral column, together with the intervertebral discs, play an important part in absorbing the shocks and stresses produced by locomotion and thereby protect the delicate tissues of the brain from damage.

Development of the skeleton of the limbs

The *pectoral* and *pelvic girdles* are formed mainly by endochondral ossification, although an important exception to this rule is the *clavicle*, which is largely intramembranous in origin. The bones of the limbs develop from the somatic mesoderm present within the limb buds. Primary ossification centres appear in the cartilagenous models of the long bones during the 8th week of development, and secondary centres appear much later, between birth and the twentieth year. After the cartilagenous models of the long bones have been formed the mesodermal cells between the ends of adjoining bones differentiate to form joints.

The Urinary System

Most parts of the urinary system develop in the intermediate mesoderm, which lies on each side of the midline between the somites and the more laterally placed somatic and visceral mesoderm, as shown in *Figure 23*. Three systems of urinary tubules appear in a craniocaudal sequence: the *pronephros*, the *mesonephros*, and the *metanephros*, the positions of which are shown in *Figure 53*. Only the metanephros develops into the definitive kidney in man. This sequence of transitory stages indicates in a general way how evolution of the urinary system as we know it has occurred. In some species the pronephros is retained as the major system for the excretion of nitrogenous wastes, while in more complex forms it is superseded by the mesonephros. In human embryos, however, the pronephros is rudimentary and probably nonfunctional, and even the mesonephros is eventually replaced in this particular role, although in male embryos it then plays an important part in the development of the reproductive system. The kidneys finally develop from the metanephros situated at the caudal end of the embryo and then migrate in a cranial direction to take up their position on the posterior abdominal wall.

The pronephric tubules of the human embryo appear briefly during the 3rd week of development. The first-formed tubules regress as the more caudally placed tubules arise, and the entire pronephros disappears by the end of the 4th week.

The mesonephros resembles the pronephros in being a series of tubules formed in the intermediate mesoderm, but it reaches a more advanced degree of differentiation and activity (*see Figure 53*). It extends throughout the thoracic and lumbar regions of the embryo, eventually forming a spindle-shaped ridge projecting into the intraembryonic coelom on each side of the mesentery supporting the midgut and hindgut. The medial end of each mesonephric tubule enlarges and becomes associated with a cluster of blood capillaries, which arise from branches of the dorsal aorta. The lateral end of the tubule opens into a longitudinal collecting duct called the *mesonephric duct* (or sometimes the *Wolffian duct*), the caudal end of which grows towards the cloaca and soon opens into it. The mesonephric tubules continue to grow, and they become elaborately curved, the more cranial ones beginning to function during the 6th week of development. They secrete copious quantities of dilute urine. However, after a short period of activity the tubules start to degenerate – again in a craniocaudal sequence – and

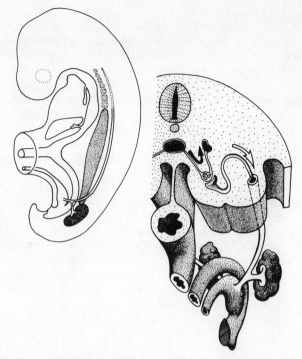

Figure 53. The developing urinary system. *Left:* The positions of the rudimentary pronephros (dotted segmental blocks), the mesonephros (medium stipple), and the metanephros (heavy stipple) are shown. The long mesonephric duct is indicated as it passes caudally to enter the allantois, and it is linked with the metanephric rudiment by the outgrowing ureteric bud. *Right:* In this reconstruction the arrangement of a mesonephric tubule is shown. Blood from the dorsal aorta passes through a capillary network, and fluid enters the mesonephric tubule. It is collected by the mesonephric duct and discharged into the allantois.

by the end of the embryonic period the majority have disappeared. Those that remain become associated with the reproductive system.

The metanephric kidneys develop from two sources: the *ureteric buds,* which grow out from the mesonephric ducts, and the *metanephrogenic caps* of intermediate mesoderm. On each side the ureteric bud arises as an outgrowth from the mesonephric duct near its opening into the cloaca. The bud grows into the intermediate mesoderm of the lumbosacral region, and this responds by condensing around it to form the metanephrogenic cap. If a ureteric bud fails to develop, the metanephrogenic cap, and hence the kidney, also fails to develop. The ureteric bud eventually gives rise to the *ureter,* which dilates at its upper end to form the funnel-shaped *pelvis of the ureter.* The pelvis gives off branches that in turn continue to divide and ramify within the metanephrogenic mesoderm, forming a system of *collecting tubules.* New collecting tubules are developed up to the end of the 5th month.

While collecting ducts are being developed from the ureteric buds, tubules are differentiating in the metanephrogenic cap. They resemble mesonephric tubules in having an expanded end — the *glomerular capsule* — that encloses a cluster of capillaries called a *glomerulus,* and the glomerular capsule is followed by a long duct. As the duct continues to grow, different regions can be identified. Adjacent to the glomerular capsule the tubule becomes twisted and tortuous to form the *proximal convoluted tubule.* This is followed by a long U-shaped portion, *Henle's loop,* which leads into another much folded section called the *distal convoluted tubule.* The distal convoluted tubule soon establishes contact with a collecting tubule derived from the ureteric bud, and at the point of contact their central lumina become continuous. In this way a functional unit of the kidney — the *nephron* — is formed. Nephrons are produced throughout the foetal period and possibly for a time after birth, until a total population of about 1 million nephrons has been established in each kidney.

Induction plays an important role during development of the nephrons. By explanting fragments of the relevant tissues into nutrient media and combining them in various ways, it has been shown that terminal portions of the ducts derived from the ureteric bud induce the formation of metanephric tubules in the mesoderm of the metanephrogenic cap. In the absence of collecting ducts, tubules do not appear. More recently it has been shown that the metanephrogenic mesoderm has a reciprocal effect on the ureteric bud, inducing the characteristic branching pattern of the duct system.

The nephrons begin to excrete dilute urine before birth. It passes down the ureters to the bladder, which is derived partially from the proximal part of the allantois, and is then discharged into the amniotic cavity. Nitrogenous wastes excreted in this way are prevented from accumulating in the amniotic fluid by a mechanism that also allows a valuable rehearsal of the actions of swallowing and digestion. Large quantities of amniotic fluid are swallowed by the foetus each day, and the water and waste products that it contains are absorbed into the bloodstream via the lining of the digestive tract. The waste products are finally eliminated through the placenta into the maternal bloodstream. Thus the excretory activity of the metanephric kidneys does not play an essential role in prenatal nitrogenous excretion; it is another example of prenatal rehearsal for postnatal function.

The metanephric kidneys begin their development in the sacral region of the embryo. As development proceeds the kidneys change their position and appear to ascend to a more cranial position in relation to the posterior abdominal wall (*see Figure 54*), finally reaching the upper lumbar region. The mechanism of this movement is not clear, but rapid growth in the lumbar and sacral regions and a change in the curvatures of the body may contribute.

Occasionally, one or both kidneys are arrested at some stage during this ascent and remain within the pelvic region, but the abnormal position may not have any adverse effect on kidney function. Ascent of the kidneys is almost completely prevented by a condition known as *horseshoe kidney*, illustrated in *Figure 55*. This is caused by fusion of the caudal poles of the developing kidneys resulting in a single U-shaped mass of kidney tissue. The interconnecting bridge becomes hooked beneath a midline branch of the aorta, which prevents further ascent. Consequently, horseshoe kidneys tend to occupy a position low in the lumbar region. An unfortunate consequence of this U-shaped configuration of kidney tissue is that the ureters become constricted as they pass over the connecting bridge, and

A Description of Normal Development

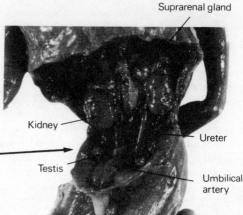

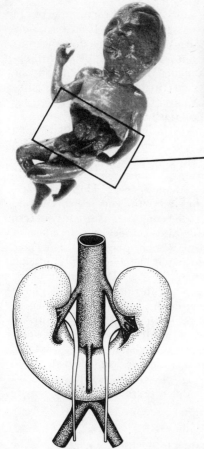

Figure 54. The urinogenital system in a 13-week human foetus. The abdomen has been dissected to show the lobulated kidneys, the ureters, and the testes. The suprarenal glands and the umbilical arteries are also visible.

Figure 55. Horseshoe kidney. Fusion of the lower poles of the kidneys prevents the ascent of the kidney tissue past the junction of the inferior mesenteric artery with the abdominal aorta. Normally the kidneys would ascend higher than this.

this restricts the flow of urine. As a result, infections and stone formation are frequent complications. *Polycystic kidney* is a congenital malformation in which the primary fault is thought to be a failure of the metanephric tubules to unite with the collecting tubules. In this condition, both kidneys are greatly enlarged and contain many urine-filled cysts, so that any functional kidney tissue becomes severely compressed and will eventually fail.

The Genital System

Development of the gonads

The genetic sex of the embryo is determined at the time of fertilization. It has been mentioned already that during formation of the gametes the number of chromosomes is halved — this applies equally to the autosomes and the sex chromosomes — so that each gamete contains twenty-two autosomes plus one sex chromosome.

In the case of the ovum the sex chromosome is always an X chromosome, but in the case of a spermatozoon it may be an X *or* a Y chromosome. If the fertilizing spermatozoon carries twenty-two autosomes and a Y chromosome, the conceptus will have a genetic complement of forty-four autosomes plus XY sex chromosomes (i.e. 44 + XY) and will be genetically male. If, on the other hand, the spermatozoon carries an X chromosome, the conceptus will be forty-four autosomes plus XX sex chromosomes (or 44 + XX) and thus female. However, for the first 6 weeks of development after conception it is not possible to distinguish between male and female embryos on morphological grounds alone, and it is only after this *'indifferent period'* that sexual dimorphism begins to appear.

It is interesting that the cells that will eventually give rise to the gametes during the individual's period of reproductive maturity are detectable at a very early stage in development — in fact, during the indifferent period. These *primordial germ cells*, as they are called, are first seen in the wall of the yolk sac close to the allantois during the 4th week (*see* Figure 45). As the embryo mushrooms out into the amniotic cavity, the primordial germ cells migrate via the dorsal mesentery of the hindgut until they reach the viscinity of the mesonephric swellings. They take up their final position medial to the mesonephros and just deep to the epithelial lining of the intraembryonic coelom, as shown in *Figure 56*. The presence of the germ cells causes the overlying epithelium to proliferate, and cords of epithelial cells grow down into the mesodermal tissue through which the germ cells are scattered. The cords of cells, which at this stage retain contact with the surface epithelium, engulf the germ cells. This burst of proliferative activity causes a ridge to form on each side of the dorsal mesentery just medial to the mesonephros. It is referred to as the *genital ridge*. At this stage it is still impossible to distinguish between the male and female gonad. During the 7th week, however, differentiation of the male gonad — the testis — diverges from that of the female gonad — the ovary.

Male

The development of the male reproductive system is illustrated in *Figure 57*. The cords of epithelial cells lose contact with the coelomic epithelium, and the

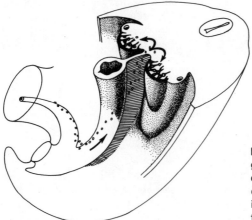

Figure 56. Formation of the primitive gonad. Arrival of the primordial germ cells initiates development of cords of epithelial cells from the lining of the intraembryonic coelom. [Modified from Langman (1975)]

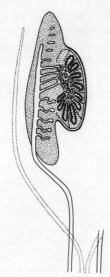

Figure 57. Development of the male reproductive system. Some of the mesonephric tubules become incorporated into the male system, and the mesonephric duct differentiates to form the epididymis and vas deferens. The paramesonephric duct (dotted lines) makes only a transient appearance.

mesodermal cells lay down a dense fibrous layer, the *tunica albuginea*, in the intervening region. Later, during the 4th month, the epithelial cords become U-shaped and form the precursors of the *seminiferous tubules*. The primordial germ cells in the seminiferous tubules form *spermatogonia*, which after puberty will take part in spermatogenesis to produce spermatozoa. The cells derived from the coelomic epithelial cells will form the supporting *Sertoli cells*. The connective tissue and hormone-secreting interstitial cells of the testis are developed from mesodermal cells. Before puberty the seminiferous tubules are composed of solid cords of cells, and they only acquire a lumen when gonadotrophic hormone from the pituitary initiates gametogenesis. The free ends of the tubules join one another in the *mediastinum* of the testis where they form a network of ducts called the *rete testis*.

Other parts of the male reproductive system are built from some of the already existing mesonephric tubules and ducts. These are relinquishing their role as an excretory system to the metanephros and many of the tubules are degenerating, but in the male embryo some of the tubules are retained as the *efferent ductules* of the testis, and the mesonephric duct is retained to carry gametes from the gonad to the urethra. The mesonephric ducts thus differentiate into the *epididymis* and the *vas deferens*. A small pocket-like outgrowth from the vas deferens close to its junction with the urethra as it leaves the bladder develops into the *seminal vesicle*, a secretory structure that adds fructose to the seminal fluid to provide an energy-giving substrate for the spermatozoa. The urethra is formed mainly by the *urogenital sinus*, one of the chambers produced when the cloaca is subdivided by the urorectal septum (*see* p. 70). The more caudal chamber develops into the rectum. A group of endodermal buds grows out from the urethra into the surrounding mesoderm, close to the points of entry of the mesonephric ducts, to form the *prostate gland*.

During development of the male reproductive tract a pair of *paramesonephric ducts* appear, but they quickly regress and disappear almost completely. These ducts run parallel to the mesonephric ducts on their lateral aspect and are formed

84 Before Birth

by the closure of a furrow-like groove in the coelomic epithelium covering the genital ridge. Thus in the male the paramesonephric ducts are transient structures with little developmental significance. However, in the female reproductive system they play a very important role.

Female

Figure 58 shows the development of the female reproductive system. If the embryo is female, the epithelial cords containing the primordial germ cells become broken up into irregular clusters during the 7th week, and the surrounding mesoderm proliferates. A thick tunica albuginea does not develop in the ovary as it does in the male gonad. The primordial germ cells mature and become *oogonia*, which during the 3rd month in particular divide mitotically to produce a large population of *primary oocytes* positioned predominantly near the periphery of the ovary. The primary oocytes are surrounded by cells derived from the epithelial cords. These cells flatten to form a thin coherent shell or *follicle* around each oocyte. Some of the follicles degenerate before birth, and this process of *atresia* continues after birth until the end of the woman's reproductive period, so that only a small proportion of the oocytes reach maturity and are ovulated.

The *uterine tubes, uterus,* and *vagina* are formed from the paramesonephric ducts (*Mullerian ducts*) and the wall of the urogenital sinus. After invagination of the coelomic epithelium to form the paramesonephric ducts, the cranial end of each duct remains open and communicates with the abdominal cavity. Finger-like projections are developed around the openings and become the *fimbria* of the uterine tubes. The caudal end of each tube grows medially and crosses ventral to the mesonephric duct until it contacts the growing tip of the paramesonephric duct from the other side. They fuse, and the single bud produced continues to elongate until it reaches the posterior wall of the urogenital sinus, usually in the 9th week of development. As the caudal ends of the paramesonephric ducts

Figure 58. Development of the female reproductive system. The ovary lies on the medial aspect of the mesonephros (light stipple) but does not utilize any of the degenerating mesonephric tubules or the mesonephric duct (dotted lines). The paramesonephric duct joins its opposite number to establish the uterus and part of the vagina. The epithelial cords containing the primordial germ cells are shown penetrating the ovarian stroma.

perform this migration from a position lateral to the mesonephric to the midline, they cause the formation of a transverse fold of coelomic epithelium between the developing rectum posteriorly and the primitive bladder anteriorly. This fold is known as the *broad ligament,* and in the adult it can be seen as a transverse partition within the pelvis, its function being to support the ovaries, the uterine tubes, and the uterus (*see Figure 68*).

The uterine tubes are formed from the cranial two-thirds of the paramesonephric ducts, and during the 4 – 5th months they elongate rapidly, becoming loosely coiled. Muscle fibres develop in the walls of the tubes, and a specialized ciliated lining is produced.

The uterus is formed from the caudal fused sections of the paramesonephric ducts. The fusion is at first incomplete, and the lumen is partially subdivided by a septum. Normally the septum disappears to leave a single cavity within the uterus.

There is still controversy about the way in which the vagina develops. One hypothesis suggests that the upper part is formed by the fused ends of the paramesonephric ducts as a continuation of the uterus, and that the lower part is formed in the wall of the urogenital sinus. The alternative idea, which is more widely accepted at present, is that the vagina is formed wholly by a solid outgrowth from the wall of the urogenital sinus, with a lumen developing secondarily in the period from 11 weeks to 5 months. If this is the case, the urogenital sinus eventually gives rise to the vagina and the urethra in female embryos.

It is possible for the development of the female reproductive system from the paramesonephric ducts and the wall of the urogenital sinus to go astray. For example, failure of the paramesonephric ducts to fuse results in partial or complete duplication of the uterus; and abnormal canalization of the vagina can occur, also resulting in duplication or perhaps obstruction.

External genitalia

Figure 59 illustrates the development of the external genitalia, both during the indifferent period and during later stages when sexual dimorphism becomes apparent. The external genitalia are formed from aggregations of mesodermal cells around the cloacal membrane. Three swellings become visible externally during the 6th week, but at this stage there is no apparent difference between male and female embryos. One swelling arises on the midline between the cloacal membrane and the umbilical cord – this is the *genital tubercle* – and two other swellings (the *genital folds*) develop one on each side of the cranial part of the cloacal membrane. In the 7th week the genital tubercle enlarges to form the *phallus,* which has an expanded end called the *glans.* Concurrently a second pair of swellings appear lateral to the genital folds, and they are known as the *genital swellings*. The part of the cloacal membrane that separates the urogenital sinus from the amniotic cavity now perforates so that the sinus opens onto the surface of the embryo.

Male
The phallus elongates rapidly, and the genital folds become pulled towards the midline, thus forming a furrow on the caudal (later the undersurface) of the phallus. The two genital folds begin to fuse, and the *penile urethra* is formed. At this

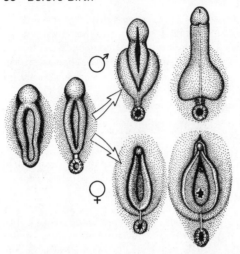

Figure 59. Development of the external genitalia. *Left:* Two stages during the indifferent period; these are the same in embryos of either sex. *Right:* Later stages during which sexual dimorphism becomes apparent (*top:* male; *bottom:* female). [Modified from Langman (1975)]

stage the urethra only extends to the root of the glans, but during the 4th month a bud of ectodermal cells from the tip of the glans grows proximally through the mesodermal core and joins the penile urethra. This cord of cells becomes canalized, so that the urethra opens at the tip of the glans.

If the genital folds fail to unite completely on the undersurface of the developing penis, or if the bud of ectodermal cells at the tip of the glans fails to contact the endodermal penile urethra, the point at which the urethra opens onto the surface will be abnormal, situated anywhere between the glans and the perineum. This abnormality is called *hypospadias*. More rarely, the urethra opens onto the dorsal or upper surface of the penis as a result of insufficient mesoderm in the anterior abdominal wall in the pubic region (epispadias).

Female

The changes in the external genitalia of the female are less marked than those in the male. The phallus does not grow rapidly in relation to other structures and becomes the *clitoris*. The genital folds do not fuse to form a continuation of the urethra but remain as folds, the *labia minora*, flanking the vestibule.

Descent of the gonads

The gonads originate on the posterior abdominal wall in a position that is relatively cranial to their definitive position. In the male, the testes leave the abdominal cavity completely and come to lie in the *scrotum*, a bag-like structure developed from the genital swellings, while in the female there is also a caudal migration although it is less marked than in the male.

Male

Descent of the testes into the scrotum ensures that they will be maintained at a slightly lower temperature than the rest of the body — an essential prerequisite for spermatogenesis — and failure of the testes to descend leads to infertility. At present there is not a complete explanation of the mechanism of descent, but it is believed that differential growth and the action of a structure called the *gubernaculum* are the two most important factors involved. The gubernaculum consists of a column of mesodermal cells extending from the caudal end of each genital ridge to the *inguinal region* (the junction between the thigh and the anterior abdominal wall) and beyond into the genital swelling.

The first stage in testicular descent seems to be the result of differential growth: the posterior body wall grows rapidly while the gubernaculum remains almost constant in length. By the 3rd month of foetal development the testis is close to the inguinal region. As this initial relative movement is occurring, an extension of the abdominal cavity enters the genital swelling, taking with it a lining of coelomic epithelium and a covering derived from each of the layers of the abdominal wall. In this way the *inguinal canal* is formed, linking the abdominal cavity with the cavity — the *processus vaginalis* — in the genital swelling. During the 7–8th months the testis passes through the inguinal canal, not within the extension of the abdominal cavity but just dorsal to it, and eventually enters the genital swelling. At the same time the two genital swellings fuse to form the scrotum. This final descent of the testis through the inguinal canal is accompanied by a shortening and thickening of the gubernaculum, but it has not been shown conclusively that the gubernaculum actively pulls the testis.

On completion of testicular descent, the connection between the abdominal cavity and the cavity within the scrotum normally becomes obliterated, leaving an isolated sac called the *tunica vaginalis* in relation to each testis. If isolation of the tunica vaginalis does not occur in this way, there is a risk of fluid from the abdominal cavity accumulating in the scrotum — a condition known as *hydrocoele* — or of herniation of loops of intestine into the scrotum. Testicular descent is completed before birth in about 97 per cent of male babies, and in many of the cases in which descent is incomplete at birth the testes enter the scrotum during the first months of postnatal life. If, however, a testis should remain within the abdomen or the inguinal canal, the higher temperature there produces irreversible damage, particularly to the seminiferous tubules. On the other hand, the interstitial cells of undescended testes can function normally, and at puberty they secrete the male sex hormone *testosterone,* with the result that male secondary sexual characteristics develop even when testicular descent has failed. Undescended testes may be positioned anywhere along the normal path of descent, or they may have migrated in a completely aberrant fashion if the gubernaculum itself is malformed.

Female

In the female, as in the male, the caudal end of the developing gonad is connected with the genital swelling by a column of mesoderm called the gubernaculum. However, as development proceeds the gubernaculum becomes attached to the developing uterus, thus subdividing it into two portions: the portion connecting the ovary to the uterus is called the *round ligament of the ovary,* and the part that extends from the uterus to the genital swelling via the inguinal canal becomes the *round ligament of the uterus.* The ovaries descend to the pelvic brim during the 3rd month. Later they take up their final position on the *broad ligament*, a partition that spans the pelvic cavity from side to side and gives support to the reproductive organs (*see Figure 68*). The attachment of the round ligament of the ovary to the uterus prevents further descent of the ovaries. Thus the early modification of the gubernaculum in the female precludes migration comparable to that occurring in the male, although there have been rare reports of migration of ovaries into the *labia majora,* which are derived from the genital swellings.

Endocrine factors in sexual dimorphism

The final pattern of sexual differentiation seems to depend on an interplay between genetic factors, particularly those related to the sex chromosomes, and hormonal factors. The hormonal stimuli may be *endogenous*, arising, for example, within the developing gonads, or *exogenous*, with maternal hormones crossing the placental barrier and influencing development. The relative significance of these factors during development is still being argued. Some forty years ago it was tentatively proposed that development will tend to follow the female pattern unless endogenously produced male hormones are present. However, it was not until the late 1940s that experimental embryologists clearly demonstrated that testosterone secreted by the foetal testes is necessary for development of the male genital ducts and external genitalia, and that female-like development occurs in the absence of the testes. More recently still, experiments in animals have shown that the maleness or femaleness of postnatal behaviour also depends on the blend of hormones circulating before birth or during the neonatal period. It appears that the functioning of the brain is 'set' by hormonal factors during a brief critical period and that from then on it responds in a predominantly male or female way.

The Foetal Period

Maturation and growth

The foetal period is characterized by two processes: *growth* and *maturation*. There is, of course, no abrupt change between the embryonic and foetal periods, but it is a useful distinction to make. For example, all the major structures and systems are established during the embryonic period, and only a few new structures appear during the foetal period. However, during the thirty weeks of the foetal period, the miniature human being grows in length by a factor of eight times and increases in weight by a factor of 600 times – figures that illustrate the phenomenal rate of growth that occurs before birth. In addition, the foetus is far less sensitive than the embryo to the teratogenic effects of certain drugs, viruses, and radiation, and any abnormalities produced tend to be of a more subtle type than those induced during the embryonic period. Many of the structures established during the embryonic period are nonfunctional or have only limited capabilities, so for them the foetal period is one of maturation and preparation to meet the needs that will arise after birth. Many essential activities are rehearsed and perfected before birth – e.g. sucking, swallowing, formation of urine – because there will be no room for mistakes once the baby is outside the security of the womb.

Some highlights of the foetal period

Table 1 summarizes the relationships between age, length, and weight of the foetus, also the main external features, between the 9th and 38th weeks of development. Further details of important changes follow, divided into significant periods.

A Description of Normal Development

Table 1. Major characteristics at different stages of the foetal period.

Age (weeks)	Crown–rump length (mm)	Weight (g)	External features
9	50	8	Comparatively large head. Midgut herniation. Wide-set eyes (lids usually closed). Low-set ears.
10	60	14	Midgut returning to abdomen. Fingernails developing.
12	87	45	Sex distinguishable externally. Forelimbs well developed.
14	120	110	Head erect.
16	140	200	Ears stand out from head.
18	160	320	Vernix caseosa present. Hindlimbs reach final proportions. Toenails developing.
20	190	460	Hair visible.
22	210	630	Skin wrinkled.
24	230	820	Lean but well-proportioned body.
26	250	1 000	Eyes partially open. (*Foetus may survive premature delivery from this time on.*)
28	270	1 300	Eyes open. Plentiful hair.
30	280	1 700	Body becoming plumper.
32	300	2 100	Smooth skin.
36	340	2 900	Lanugo hair is shed.
38	360	3 500	Prominent chest. Testes in scrotum or inguinal canals. Fingernails extend beyond fingertips.

9–12 weeks

The proportions of the foetus begin to change as different parts of the body grow at different rates. Initially the head is as large as the trunk, but now the rest of the body begins to grow more rapidly, and by 12 weeks the head is only two-thirds of the size of the trunk. The face is broad, the eyes are widely separated, and the ears are low-set. The upper limbs are growing rapidly and by the end of this period have almost attained their final proportionate length, but the lower limbs are lagging behind at this stage. The external genitalia become recognizably male or female by the middle of this four-week period, and the normal umbilical herniation of the midgut regresses at about the same time. During this period the foetus already shows signs of activity, including the sucking reflex, although it will be several weeks before the mother becomes aware of foetal movement. *Figures 60 and 61* show a 10-week and 12-week foetus respectively.

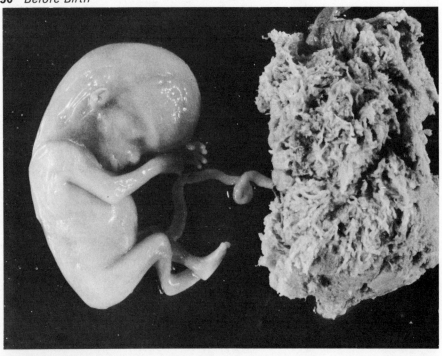

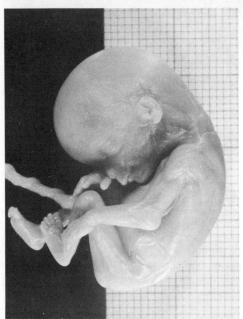

Figure 60. A 10-week foetus and placenta.

Figure 61. A 12-week foetus.

13–16 weeks

Growth of the body is very rapid during this period. The head continues to grow relatively slowly, however, and by 16 weeks it is half as long as the trunk. Ossification of the skeleton, a process that began at the end of the embryonic period, has now become sufficiently advanced to show clearly on X-ray films.

17–20 weeks

The rate of growth begins to slow during this period, but even so the length increases by about 40 mm to become 190 mm at 20 weeks, and the weight increases by about 200 g to become 460 g. The lower limbs reach their final relative proportions. The mother begins to notice the foetal movements. The skin of the foetus is covered by a protective layer of a waxy material called *vernix caseosa*, which, it is thought, limits abrasion of the skin as the foetus moves and protects the skin against the urates present in the amniotic fluid. At 20 weeks foetuses are usually completely covered with fine downy hair called *lanugo*. The coarser eyebrows and head hair also develop at this time. A 20-week foetus is shown in *Figure 62*.

21–25 weeks

The increase in weight continues to be substantial — about 350 g during these weeks — and the proportions of the body continue to change, although less rapidly and more subtly than before. All the organs and systems are quite well developed, but a foetus born at this stage usually dies as a result of respiratory difficulties (*See Figure 13*, which illustrates a 6-month foetus.).

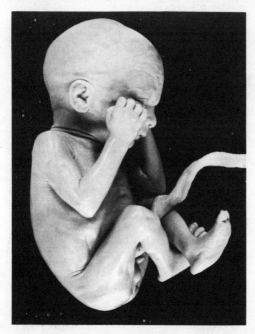

Figure 62. A 20-week foetus.

26–29 weeks

The prospects of survival are considerably better if premature birth occurs in the 26th week or later. The central nervous system has matured to the stage where it can control respiratory movements and maintain a stable body temperature. However, the mortality rate in premature infants is high, again largely as a result of respiratory difficulties. The contours of the foetus become smooth and rounded as a layer of subcutaneous fat is established beneath the skin.

30–38 weeks

During the 'finishing' period the rate of growth slows considerably. By the time of birth the foetus usually appears quite chubby and weighs about 3 500 g. However, this final weight is influenced by a number of factors. For example, the growth rate is retarded if the mother has a poor diet or if she smokes cigarettes. Similarly, twins, triplets, etc. usually weigh far less than infants resulting from a single pregnancy. On the other hand, infants of diabetic mothers tend to be larger than normal.

Birth generally occurs about 280 days after the onset of the last menstrual period before pregnancy. Thus, if it is assumed that ovulation, and hence fertilization, occur about fourteen days later, this means that prenatal development normally takes about 266 days, or thirty-eight weeks. Most babies are born within ten to fifteen days on either side of this time. It has already been noted that very premature birth is dangerous for the foetus, but postmaturity can also give problems because the placenta rapidly becomes inadequate as term approaches. If birth is delayed significantly the foetus begins to lose weight, his skin becomes parchment-like, and he may begin to suffer from lack of oxygen.

The Amnion, Chorion, and Placenta

It is now necessary to retrace our steps along the path of development and return to the time of implantation in order to appreciate the changes that occur in the *extraembryonic membranes* as the embryo and foetus develop. Strictly speaking, a description of the developmental fates of the distal parts of the allantois and the yolk sac should be included in this account as they are not incorporated into the embryonic body. However, their development has been touched upon already (*see* pp. 61 and 80), and therefore this subsection will be confined to a description of the amnion and chorion. Most emphasis will be given to a specialized region of the chorion that, with the co-operation of maternal tissues, forms the *placenta*.

The amnion

The definitive amnion is a thin, tough, transparent tissue that forms the wall of the amniotic cavity. Initially it consists of a single layer of cells continuous with the embryonic ectoderm around the periphery of the embryonic disc. (*See Figure 6.*) As the embryo grows and folds, it mushrooms more and more into the enlarging amniotic cavity, and the junction with the amnion becomes narrowed and shifted to a ventral position on the body wall. In this way the connecting stalk between the embryo and the trophoblastic shell becomes invested totally by amnion. The stalk contains the distal extensions of the yolk sac and allantois, together with blood vessels linking the embryo to the developing placenta, and will ultimately become the *umbilical cord*.

The mature amnion is composed of two layers: an inner epithelial layer derived from embryonic ectoderm, and an outer layer of extraembryonic mesoderm. The amniotic cavity is filled with a pale watery fluid, the *amniotic fluid,* which increases in quantity until the 6th or 7th month of pregnancy and then diminishes a little. At term, it normally amounts to approximately 1 l. in volume. In the early stages the fluid is produced by the amnion, but later, when the foetal kidneys become functional, urine is added to the fluid and becomes a major component. The amniotic fluid is prevented from stagnating by a remarkable mechanism: from about the 4th month onwards the foetus swallows large quantities of the fluid. Most of it is absorbed by the digestive tract and then transported by the foetal blood to the placenta, where excess fluid and waste materials are passed into the maternal bloodstream, while a much smaller proportion may be drawn into the foetal respiratory system. It has been estimated that towards the end of pregnancy there is a turnover of one-third of the total water content of the fluid every hour by this process of swallowing and absorption.

There are conditions in which the volume of amniotic fluid is either too great (*hydramnios*) or too little (*oligohydramnios*). In hydramnios, as much as 6 l. of fluid may be present. It is commonly associated with maternal diabetes but may also be the result of an inability of the foetus to swallow and absorb the fluid in the normal way. Thus, severe derangements of the brain, such as anencephalus, or blockage of the digestive tract predisposes towards hydramnios. Oligohydramnios is marked by an insufficient quantity of fluid, usually indicating poor functioning of the foetal kidneys, and may result in deformities of the foetus due to intense localized uterine pressure.

Amniotic fluid has several functions: it provides the foetus with a protective cushion that reduces external trauma; it provides a medium within which the foetus can move freely; it helps to maintain a stable temperature; and it plays an important role during the first stage of labour, not only by protecting the foetus against the powerful uterine contractions, but also by forcing a wedge of chorion and amnion into the uterine cervix and so dilating it. This hydrostatic wedge advances down the cervical canal with successive contractions until the membranes burst. When this happens some of the amniotic fluid escapes, and this event is often referred to as the appearance of the 'waters'.

Samples of amniotic fluid can be extracted from the amniotic cavity with comparative safety from about the 3rd month of pregnancy onwards. The fluid contains cells sloughed from the foetus, and tests can reveal the sex of the foetus or indicate the absence of essential enzymes or warn of the presence of chromosomal abnormalities. A further application of this technique of *amniocentesis* is given in the section on abnormal development (*see* p. 123).

The chorion

The chorion consists of the trophoblastic shell together with a lining of extraembryonic mesoderm. Just before implantation, the trophoblast begins to change its configuration as a single layer of cells arranged in a hollow sphere and becomes a bilaminar structure (*see Figure 5*). The internal layer is composed of normally organized cells, while the external layer consists of a syncytial mixture of nuclei and cytoplasm that shows no evidence of separation into discrete cellular units. The two layers of the trophoblast are therefore referred to as the *cytotrophoblast* and the *syncytiotrophoblast*.

As implantation nears completion, cavities called *lacunae* appear in the syncytiotrophoblast and in the adjacent uterine endometrium (*see Figure 6*). The endometrial cells enlarge and become filled with stores of glycogen and lipids. These changes spread away from the implantation site to include eventually the whole lining of the uterine cavity. The modified endometrium is now known as the *decidua*, an appropriate term in view of the way in which much of this tissue is shed at birth with the foetal membranes. The enlarging lacunae of the trophoblast and endometrium coalesce, and the congested uterine capillaries near the implantation site discharge maternal blood into them. Initially there is accumulation of blood without circulation, but later a definite circulation past the trophoblastic tissue is established, with the blood supplied by maternal arterioles being taken up by maternal venules. Thus, soon after implantation an intimate relationship is established between the maternal blood and the outer surface of the conceptus.

Trophoblastic villi

During the period from 9 days to 20 days after conception, the trophoblastic shell becomes highly specialized and expands rapidly. Finger-like cords of cytotrophoblastic cells migrate centrifugally into the irregular syncytial layer to produce composite structures called *primary villi*. Each villus has a central core of cytotrophoblast and an outer covering of syncytiotrophoblast and extends into a region between adjacent blood-filled lacunae. Extraembryonic mesodermal cells, which are developed largely from the internal aspect of the cytotrophoblast, now invade the cytotrophoblastic core of each villus and convert it into *secondary villi*. The cells at the tips of most of the villi proliferate rapidly until each blood-filled lacuna is fully lined by trophoblastic tissue.

Networks of blood capillaries begin to differentiate in the cores of the secondary villi and in the layer of extraembryonic mesoderm lining the trophoblastic shell. This process occurs concurrently with the similar formation of blood vessels in the embryo and body stalk, and subsequently the intraembryonic and extraembryonic networks become linked to form a single complex. At about the same time, soon after the end of the 3rd week of development, the heart begins to beat and a primitive circulation of blood is established within the conceptus (*see Figure 32*). Blood is pumped from the embryo through the body stalk to the capillaries in the secondary villi, and it then returns, again via the body stalk, to the embryo.

Initially, the outer surface of the chorion is evenly covered by secondary villi, as shown in *Figure 9*, although those situated on the more deeply implanted aspect of the conceptus are larger than those developed more superficially. As the conceptus enlarges, the roof of endometrium over the implantation site begins to bulge into the uterine cavity. Thus the superficial layer of decidua becomes increasingly stretched and thinned, with the result that the secondary villi in that region begin to regress until eventually they disappear, leaving part of the chorion devoid of villi. Conversely, the villi developed from the opposite deep pole of the conceptus become larger and more elaborate. Usually the body stalk links the embryo to the centre of this region of elaboration of the chorion.

By the 4th month of development the conceptus has enlarged so much that the uterine cavity is completely obliterated and the thin layer of decidua covering the original implantation site fuses with the decidua lining the opposite wall of the uterus. The uterine cervix does not undergo the decidual reaction, but its glands enlarge and secrete mucus, which blocks the cervical canal. At the end of pregnancy this *cervical plug* is pushed out by the hydrostatic wedge of amnion and chorion during the first contractions of labour.

The placenta

Structure of the placenta

The placenta is composed of foetal and maternal tissues. The foetal part consists of the deeply implanted pole of the chorion with its outgrowth of elaborate secondary villi, and the maternal part consists of the nearby decidual tissue. *Figure 63* shows the general organization of the placenta.

The growing secondary villi send numerous branches into the lakes of maternal blood circulating between the villi. At first each villus retains its two-layered covering, but during the 4th month the cytotrophoblastic cells decrease in number until their original appearance as a consistent layer is lost and only small clumps of cytotrophoblastic cells remain. Thus, in a mature branching villus the foetal capillaries are separated from the maternal blood by a thin layer of syncytiotrophoblast only. This placental membrane separating the two bloodstreams is very thin — about 0.002 mm thick — and it allows a ready exchange of materials to occur. The permeability of the membrane reaches a maximum during the 36th week, but there is a rapid decline in efficiency towards the end of pregnancy when fibrous deposits accumulate at the interface. The free surface of the syncytiotrophoblast is covered by *microvilli*, which greatly increase the surface area available for the diffusion of smaller molecules, such as gases in solution, amino acids, sugars, hormones, some vitamins, and essential inorganic ions. Certain drugs can also diffuse freely across the membrane: e.g. antibiotics, nicotine, alcohol, and general anaesthetics. However, diffusion is not the only means of placental transfer: larger molecules, in particular proteins, are actively taken up by the cytoplasm of the syncytiotrophoblast and conveyed to the foetal bloodstream. Other essential factors (e.g. vitamin C) are also actively transported across the placenta, with the result that the concentration in the foetal blood is higher than that in the maternal blood.

Foetal blood is carried to and from the capillaries in the placental villi by radially arranged vessels that converge on the point of attachment of the umbilical cord to the placenta. Here the placental vessels link with the umbilical vessels, which bring deoxygenated blood from the foetus and return the oxygenated enriched blood to the foetal heart.

Thin partitions of maternal and foetal tissue grow into the intervillous spaces and partially subdivide them, so that each villus stands within a dome-shaped chamber through which maternal blood is continuously passing. These *placental septae* separate the circular placental field into about fifteen to thirty territories, or *cotyledons*, each of which usually contains a single villous tree. The total volume of maternal blood present in the intervillous spaces of a mature placenta at any one time is about 175 ml, while the rate of turnover of blood in the spaces is about 500 ml per minute. The mild uterine contractions that occur periodically throughout

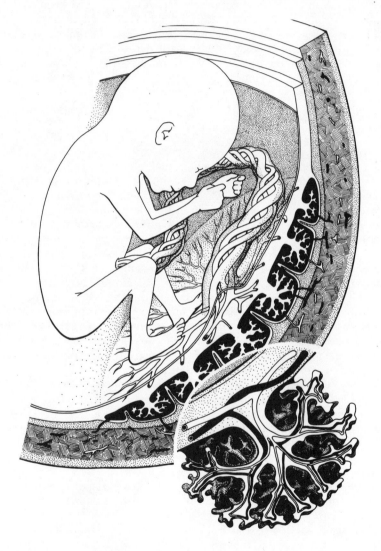

Figure 63. General organization of the placenta. Where the placenta is shown in section, the uterine components are stippled and cross-hatched, while the foetal tissue is unshaded. The lacunae of maternal blood are solid black. *Lower right:* A single villous tree in schematic form.

pregnancy aid the drainage of blood from the intervillous spaces by compressing them and forcing the blood into the uterine venules and veins. On the other side of the placental membrane, the volume of foetal blood flowing through the villi is estimated at 400 ml per minute. It is worth emphasizing at this point that, under normal conditions, *there is no mixing of foetal and maternal blood*.

The placenta at term

The placenta increases in size and thickness throughout pregnancy, but it retains its general proportions in relation to the internal aspect of the expanding uterus, covering about one-third of the foetomaternal interface. By the time of birth it is saucer-shaped and has a diameter of about 20 cm. It is thickest in the central region – about 2.5 cm – and weighs approximately 500 g. It is continuous around its periphery with the remainder of the chorion. The foetus is attached to the placenta by the umbilical cord, which is twisted and tortuous, about 2 cm in diameter and some 50–60 cm long. A few minutes after delivery of the baby the placenta separates from the uterine wall and is expelled by the last few contractions of the uterus.

The outer (*maternal*) surface of the freshly delivered placenta is covered by dark red maternal decidual tissue and has a spongy consistency. The irregular outlines of the cotyledons can be distinguished. The inner (*foetal*) surface is relatively smooth and glossy since it is covered by amnion. The larger blood vessels linking the umbilical cord with the villous trees can be seen through the amnion.

Functions of the placenta

Although the placenta has a fairly simple structural organization it performs a remarkable number of vital functions.

1. *Respiratory function.* The oxygen required during development is obtained by diffusion from maternal blood to foetal blood, through the placental membrane. The special haemoglobin carried by deoxygenated foetal red blood cells has a high affinity for the oxygen transported by maternal arterial blood, and a transfer occurs. At the same time the carbon dioxide resulting from foetal metabolism passes readily in the other direction and is carried away by the maternal bloodstream.
2. *Nutritional function.* The raw materials for differentiation and growth pass by diffusion or are actively transported across the placental membrane into the foetal blood. These materials include water, inorganic salts, carbohydrates, fats, proteins, and vitamins.
3. *Excretory function.* In addition to carbon dioxide, other waste products of foetal origin (e.g. urea) can be excreted through the placenta into the maternal blood.
4. *Protective function.* The placental membrane, while carrying out efficient transfer of materials, can also act as an effective barrier to some potentially harmful materials. In particular, many bacteria are unable to cross the placenta from the mother to the foetus. However, the barrier is not totally effective and some harmful factors, especially low molecular-weight drugs, can enter the foetal bloodstream and cause damage to the foetus.
5. *Hormonal function.* The placenta synthesizes several hormones that are essential for the maintenance of pregnancy; without them the uterus would not support further development. An important example is *placental progesterone*. The

98 Before Birth

output of maternal progesterone falls rapidly during the 3rd month as the corpus luteum regresses, and the placenta must compensate for this loss by secreting progesterone to maintain a suitable uterine environment. Other hormones secreted in large amounts by the placenta include *oestrogen* and *gonadotrophin*. Research suggests that the steroid hormones — oestrogen and progesterone — are produced by the syncytiotrophoblast, while gonadotrophin is produced by the cytotrophoblast.

Abnormalities of the placenta

Some variations in the size or shape of the placenta may produce clinical complications, and several of the more common abnormalities are shown in *Figure 64*. The positioning of the placenta in relation to the uterus is also critical. Normally it is situated in the upper half of the uterus, but if implantation occurs much below this there is a high risk of haemorrhage during pregnancy, usually after the 4th month as the foetus begins to grow rapidly. This condition is known as *placenta praevia* and may endanger both the mother and the foetus.

Over 50 per cent of the placentas delivered at the completion of normal pregnancies show regions of degeneration called *infarcts*. These are produced when maternal blood coagulates (perhaps as a result of trauma) around a villous tree and causes degeneration of that region of the placenta. At first the infarct is red, but in time the degenerate area may become replaced by white fibrous tissue. A mild degree of infarction can be tolerated and is often thought of as a normal process of placental ageing, but if the infarction is extensive the placenta becomes inadequate and the foetus may die.

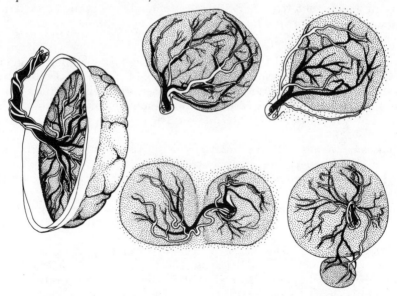

Figure 64. Different placental forms. *Left:* The most common type of placenta. *Right:* Four variants: (1) marginal attachment of the umbilical cord (battledore placenta); (2) velamentous attachment of the umbilical cord; (3) bilobed placenta; and (4) accessory lobule (succenturate placenta). Arterial branches are shown in white, venous in black.

Birth and the Neonatal Period

Birth

The mother 'goes into labour' about 9 months after the baby is conceived, but the factors that initiate labour are not clearly understood. Contractions of the uterine musculature occur weakly and painlessly throughout pregnancy, but when labour begins they become forceful and painful. The limited activity of the smooth muscle before labour may be due to an inhibitory effect exerted by progesterone, which is secreted in increasing amounts by the placenta as pregnancy progresses. The placenta itself begins to show signs of inability to meet foetal demands as term approaches — a phenomenon called *placental insufficiency* — and this may in some way trigger labour.

Another factor is the hormone — *oxytocin* — that is secreted by the posterior lobe of the pituitary gland. The uterine musculature becomes very sensitive to oxytocin late in pregnancy, and it seems that, once labour is initiated, stimuli from the uterus cause reflex secretion of the hormone, which in turn augments uterine contractions.

The pain experienced during powerful labour contractions is probably caused by anoxia in the uterine muscle; as the muscle contracts, the supply of blood becomes restricted just when the demand for oxygen is greatest. The contractions can occur in the absence of neural control, so it is possible to reduce labour pains by spinal anaesthesia while delivery continues normally.

At term, the foetus usually lies head down with his long axis parallel to the long axis of the uterus, so that the head is usually the first part of the baby to emerge (*see Figure 65*). Sometimes, however, the buttocks present first — a *breech presentation* — and this can complicate delivery.

The first stage of labour begins with the onset of regular forceful uterine contractions and is completed when the cervix of the uterus becomes fully dilated. The second stage consists of complete delivery of the baby through the vagina.

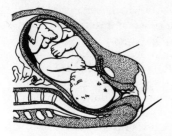

Figure 65. Delivery of the baby. The placenta is shown in solid black. [Modified from Blechschmidt (1961)]

This difficult phase is aided by a hormonally induced loosening of pelvic ligaments, particularly the pubic symphysis, and by the ability of the bones of the baby's skull to overlap and so reduce the diameter of the head as it passes along the 'birth canal'. The third stage of labour consists of the expulsion of the placenta and attached membranes. This is completed soon after delivery of the baby, requiring a few final contractions of the uterine muscle. As the placenta separates from the uterus, taking with it most of the maternal decidual tissue, many of the uterine blood vessels are torn. However, following separation, the muscular wall of the uterus contracts strongly and this constricts the vessels, thereby limiting blood loss. A careful examination of the placenta is carried out, partly to assess its normality but also to ensure that no fragments remain in the uterus. This is to reduce the risk of residual trophoblastic cells developing into malignant, highly invasive cancer cells (*chorioepithelioma*).

Changes at birth

When the baby emerges from its warm protected environment into the hostile world, those of his needs that were previously satisfied by the placenta must now be fulfilled by more active behaviour on his part. No longer is his the privileged parasitic mode of life; he must somehow overcome the problems of obtaining oxygen and nutrients, find ways of excreting waste products, and learn to maintain a stable internal temperature despite external fluctuations. His immune-response mechanism must learn to recognize and overcome invading organisms. Clearly, the time of birth and the first critical weeks of the *neonatal period* that follow are times of difficult physiological adaptation for the baby — times of great risk when adjustments must occur rapidly in a number of systems.

Respiratory system

As soon as the placental circulation begins to fall during labour, the respiratory system must effectively take over the responsibility of obtaining oxygen and eliminating carbon dioxide. Before birth the alveoli and conducting pathways of the respiratory system are filled with fluid, the source of which is unclear. It was at one time thought to be amniotic fluid sucked in during respiratory movements *in utero*. However, a recent study in which radiopaque dye was injected into the amniotic fluid casts doubt on this, since no dye could be seen within the foetal lungs. It is possible that the fluid is secreted by the lung tissue itself. Much of the fluid is ejected during birth: as the thorax passes through the birth canal it is compressed, partially emptying the lungs. The remaining fluid is probably absorbed into the pulmonary branches of the lymphatic system.

The first breath occurs within seconds of delivery and is prompted by a strong contraction of the diaphragm. Soon after this the baby gives his first cry. The initial gasps are superseded by a more relaxed and rhythmical pattern of breathing, with about forty breaths per minute. An interesting substance called *surfactant* plays a vital role at this time by contributing to the efficiency of gaseous exchange. It has detergent-like properties and acts by reducing the surface tension of the fluid film lining the alveoli. Thus it facilitates the interchange of molecules between the fluid and the air in the alveoli. Absence of an adequate quantity of surfactant can result in respiratory distress. Most babies are slightly cyanotic at birth, but normally the oxygen deficiency is soon corrected.

Circulatory system

Another system that has to make rapid adjustments during and after birth is the circulatory system. *Figure 38* shows the pattern of circulation before birth. It will be remembered that, in addition to the placental extension, the foetal circulation is characterized by a number of shunts, or short cuts. Thus blood returning from the placenta, via the umbilical vein, is shunted through the liver tissue by the *ductus venosus*. Much of the blood in the right atrium is shunted directly to the left atrium through the *foramen ovale*. Most of the blood leaving the right ventricle is channelled from the pulmonary circulation into the systemic circulation through the *ductus arteriosus*.

Each of these arrangements must be changed at, or soon after, birth. The sequence is begun as the umbilical cord is tied within a minute or so of birth. Even if the cord is not tied the placental circulation soon ceases, as the umbilical vessels spontaneously contract and limit blood flow. This abrupt halt in the return of blood from the placenta results in an immediate fall in pressure in the inferior vena cava and right atrium. At the same time, as respiration is established the pulmonary blood vessels dilate, and the resulting drop in resistance to blood flow allows a rich pulmonary circulation to develop. This means that the left atrium begins to receive an increased quantity of blood through the pulmonary veins, with a corresponding rise in pressure. Therefore the pressure difference between the right and left atria is now the reverse of that existing during foetal life, and this causes the foramen ovale to close. Usually this simple apposition is followed later by a process of fusion, so that the atria become permanently separated.

Failure of the foramen ovale to close after birth would result in mixing of blood between the right and left sides of the heart. Although mixing is essential during foetal development, postnatally it reduces the level of oxygenation of blood in the systemic arterial system. However, it is quite common to find incomplete fusion of the flaps of the foramen ovale: in about 20 per cent of otherwise normal healthy people it is possible to pass a small probe through the interatrial septum. This is known as *'probe patency'* of the foramen ovale.

The development of a flourishing pulmonary circulation is aided by constriction and closure of the ductus arteriosus. This is achieved by spontaneous contraction of its muscular wall (possibly in response to the improved oxygenation of aortic blood) and its subsequent conversion into a ligamentous structure, the *ligamentum arteriosum*. Failure of this process results in the continued mixing of pulmonary and systemic blood with consequent cyanosis. (Similarly, the ductus venosus, which has no further use when the umbilical cord is tied, is converted into the *ligamentum venosum*.)

When these changes are complete the blood circulates as in *Figure 66*.

The blood itself changes in the neonatal period. Foetal red blood cells (*erythrocytes*) are larger than those developed after birth and contain a special type of haemoglobin that is able to function adequately at the lower oxygen tensions found in the foetal circulation. During the first month after birth there is a steady decrease in circulating erythrocytes, and all infants develop a mild form of anaemia by the end of this period. However, production of the postnatal type of erythrocyte steadily rises, and there is a spontaneous recovery from the anaemia.

Digestive system

The digestive system begins to rehearse its postnatal role before birth. From

102 Before Birth

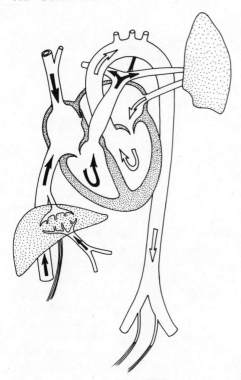

Figure 66. The pattern of blood circulation in the adult. Black arrows indicate deoxygenated blood, and white arrows indicate oxygenated blood.

about 4 months after conception the foetus swallows amniotic fluid. The sloughed-off cells and debris suspended in the fluid are attacked by enzymes secreted by the digestive system. The undigested residue accumulates in the hindgut as faecal material called *meconium*, which is defaecated during the first two or three days after birth. It seems that most of the digestive fluids and enzymes that can be identified postnatally are also synthesized before birth. An exception is pancreatic amylase; this enzyme, which breaks down starch, is produced only postnatally.

Within twenty-four hours of birth the baby begins to show signs of hunger by becoming restless and sucking on his fingers or anything else close to his mouth. His sense of smell is acute, and he reacts rapidly to the smell of milk, turning his head towards the source of the smell. Sucking requires the combined action of tongue, cheek, and palate, with the lips playing a relatively minor role. For this reason, cleft palate causes more feeding problems than cleft lip.

Urinary system

The urinary system also becomes functional before birth, and the kidneys excrete dilute urine from about the 9th week after conception. It is acid and contains few sodium, chloride, or phosphate ions. The urine is passed into the amniotic fluid. When the foetus swallows amniotic fluid, the urates it contains are absorbed from the digestive tract and finally removed from the foetal bloodstream at the placenta. Normally, the kidneys are functioning adequately by the time of birth to cope with postnatal demands. This is provided that the demands are not excessive, as could be the case when intravenous therapy is used neonatally.

Umbilical cord

The remaining stump of the umbilical cord shrivels and dries, and is shed about a week after birth. The intra-abdominal portions of the umbilical vessels become converted into ligaments, which tend to pull the overlying skin inwards to form the typical umbilicus.

Behaviour of the newborn

A newborn baby can flex and extend his limbs, yawn, sneeze, cough, and suck. He is easily startled by noise, bright light, or sudden change in posture. He sleeps for long periods until woken by hunger or discomfort. He can grasp objects firmly in his hands, and his toes also make grasping movements when pressure is applied to the sole of the foot. Most of the motor activities shown by the newborn have been established and practised *in utero*, and they are usually reflex responses that do not require control by higher centres of the central nervous hierarchy. As the nervous system matures after birth, this reflex activity is progressively suppressed by the activity of the brain and replaced by increasingly subtle and intricate types of movement.

Eye movements are at first uncoordinated, and it takes several months before images can be focused on the retina. A foetus and newborn baby respond to sounds, but the sensitivity of hearing improves greatly during the first part of the neonatal period as fluid drains from the cavity of the middle ear.

Perhaps crying is the baby's best-known skill — note that no tears are formed until the end of the neonatal period when the lacrimal glands begin to function — but it seems a little unfair to describe a baby, as an harassed father is once supposed to have done, as 'an alimentary canal with a loud noise at one end and no sense of responsibility at the other'!

Gametogenesis

The gametes produced during the reproductive period of an individual are the culmination of two processes: one a sequence of structural change and reorganization, the other a process of reduction of the chromosomal number to half the normal number. The structural preparations producing spermatozoa are very different from those producing ova, but the pattern of accompanying genetic changes is comparable in both sexes providing that no account is taken of the very different time scales involved.

Spermatogenesis

This term is applied to the sequence of changes that occurs when primitive spermatogonia in the testes are converted into spermatozoa. Spermatogenesis begins when the boy is about 14 years old and continues from this time into old age.

It will be recalled that the testes originate in the abdomen and descend before birth, usually by the end of the 8th month, to take up their final position in the scrotum. This is an important migration since spermatogenesis can only occur at a temperature lower than that existing in the abdomen. In undescended, or *cryptorchid*, testes the seminiferous tubules undergo irreversible degenerative changes.

Each testis is subdivided by fibrous septa into 250 or so separate compartments, each of which contains one or several seminiferous tubules. Interstitial cells lying among the tubules secrete the male sex hormone testosterone. The seminiferous tubules open into a network of pathways called the *rete testis*, which in turn drains via efferent ductules into a long, closely coiled duct called the *epididymis* situated on the outer surface of the testis. The epididymis consists of a duct some 5 m long and is continuous with the thick-walled *vas deferens*, which leaves the scrotum and enters the abdominal cavity. Close to the posterior wall of the bladder the vas deferens joins the duct of the *seminal vesicle* to form the *ejaculatory duct*, and this in turn opens into the *urethra*.

The wall of a seminiferous tubule is composed of a number of irregular layers of cells. Large pyramidal *Sertoli cells* extend from the external basal lamina of the tubule to the lumen, and lying between them, along the basal lamina, there are numerous germinal cells, the *spermatogonia*. Spermatogonia are of two types: type A and type B. Type A spermatogonia are stem cells, each of which divides mitotically (i.e. retaining the original chromosomal number) to produce another type A cell plus a type B cell. Thus the number of stem cells remains constant. Type B spermatogonia divide, again mitotically, to form *primary spermatocytes*. These migrate towards the middle zone of the epithelium and undergo the *first meiotic division*, during which the number of chromosomes per cell is halved. The smaller *secondary spermatocytes* produced by this division quickly divide once more to form *spermatids*, which become embedded in the cytoplasm of the luminal ends of the Sertoli cells.

The spermatids undergo a number of morphological changes that convert them into spermatozoa, as shown in *Figure 67*. The nucleus of each spermatid condenses and becomes slightly flattened and elongated. The membrane-bound vesicles of the Golgi apparatus coalesce to form an *acrosomal cap* round one end of the nucleus, and in this way the *head* of the spermatozoon is formed. Enzymes stored within the acrosomal cap are thought to play an important role during fertilization, possibly helping the spermatozoon to penetrate the zona pellucida around the ovum. Two small cytoplasmic inclusions called *centrioles* move to the pole of the nucleus opposite to the acrosomal pole. Each centriole resembles the

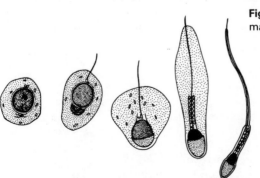

Figure 67. Transformation of a spermatid *(left)* into a spermatozoon *(right)*.

base of a cilium and consists of a geometric array of microtubules. One of the centrioles acts as a seeding point from which the motile *tail* or flagellum of the spermatozoon develops; the other centriole lies nearby but does not participate in formation of the tail. Mitochondria migrate towards the flagellum and form a spiral sheath around its proximal part, and together they are known as the *body* or middle piece of the spermatozoon. Most of the cytoplasm of the original spermatid is cast off and degenerates.

Fully formed spermatozoa are released from the Sertoli cells and become free within the lumen of the seminiferous tubule. They are carried passively to the epididymis by rippling contractions of the muscle fibres in the walls of the various ducts, and there they begin to show signs of motility.

Spermatogenesis requires about sixty-four days for completion. Not all the seminiferous tubules are at the same stage in the cycle at any one time, and varying degrees of activity can be detected along the length of a particular tubule. Abnormal spermatozoa occur quite commonly — some with two heads or two tails, others with unusual proportions — but provided they do not account for more than about 20 per cent of the total there is unlikely to be a loss of fertility.

Oogenesis

The process of oogenesis, illustrated in *Figure 68*, is the female counterpart of spermatogenesis and consists of the transformation of *oogonia* into *ova*. However, it differs from the sequence in the male in that it begins before puberty — in fact,

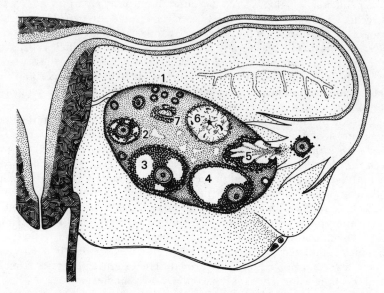

Figure 68. Stages in oogenesis. The female reproductive system is shown in schematic form, and the following structures are indicated: (1) primordial follicles; (2, 3, 4) maturing Graafian follicles; (5) ruptured follicle and ovulated oocyte; (6) corpus luteum; and (7) atretic follicle. The ovary is supported by the broad ligament, a membranous sheet extending laterally from the uterus and uterine tubes.

before birth. The oogonia begin to develop into *primary oocytes* during the 3rd month after conception, and by the time of birth they have commenced the first meiotic division. This division is not completed immediately, however, and the cells enter a resting phase. The primary oocytes are positioned in the peripheral cortex of the ovary, and a single layer of flattened cells becomes established around each of them, transforming them into *primary follicles*. Many of the follicles degenerate prenatally — at birth there are about 700 000 follicles present in the two ovaries — and further degeneration occurs after birth. Only about 40 000 primordial follicles survive to puberty.

At puberty the cyclical production of mature ova begins. The hormonal stimulus for the initiation and maintenance of this process comes from the small pituitary gland situated beneath the brain. During each four-week cycle several follicles — perhaps six to eight — begin to enlarge and mature in the ovaries, but usually only one follicle completes the process while the remainder drop out of the race, becoming degenerate (atretic). It has been suggested that this represents a form of selection of the most viable ovum from the original group of follicles. If it is assumed that one ovum is ovulated during each monthly cycle, and that a woman's reproductive period extends from about 15 to 45 years of age, it can be seen that only about 400 mature ova are liberated while the remaining great majority of follicles become atretic.

The primordial follicles begin to mature in response to follicle-stimulating hormone (FSH) from the pituitary. The follicle cells increase in number until they form a multilayered capsule around each oocyte. A glycoprotein covering — the zona pellucida — is then secreted around the steadily enlarging oocyte by the follicle cells, which also take on the task of secreting oestrogen. Irregular fluid-filled spaces appear between the follicle cells and coalesce to form a single cavity, the *follicular antrum*. The oocyte, still covered by a mound of follicle cells, projects into the antrum at one end of the follicle. The follicle is now called a *Graafian follicle*, and by mid cycle (i.e. after about fourteen days' growth) it is about 1 cm in diameter. The position of a mature follicle is marked by a distinct bulge on the surface of the ovary and is clearly visible to the unaided eye.

The meiotic division that began in the oocyte before birth is completed a few hours before ovulation, and two cells, very unequal in size, are formed. The *secondary oocyte* is the larger of the two, while the tiny *first polar body* lies compressed between the oocyte and the zona pellucida. Ovulation occurs when the pressure of the fluid that has accumulated in the follicular antrum is sufficient to rupture the follicle and the thin outer covering of the ovary. The secondary oocyte and first polar body, still contained within the zona pellucida to which a few follicular cells may adhere, are ejected quite forcibly into the peritoneal cavity. Immediately after ovulation the secondary oocyte begins the second meiotic division, which is only completed if fertilization occurs. As a result of this division a mature ovum and a second polar body are formed. The original polar body may also divide at this time to form two more secondary polar bodies, so it is not unusual to see an ovum accompanied by three secondary polar bodies. The subsequent role of the polar bodies, if any, is unknown, and it is assumed that they soon degenerate and disappear. However, it is interesting to remember that these haploid cells contain genetic material derived from the mother only, unlike the embryo, which contains a blend of material from maternal and paternal gametes. Consequently, polar bodies would not be recognized as 'foreign' by the mother.

Although there is no supporting evidence as yet, it is tempting to wonder whether this property of similarity is used by the implanting embryo as a means of persuading the uterus not to reject the conceptus.

The corpus luteum

After ovulation the walls of the ruptured follicle collapse and fold in on themselves, but the cells remain active, taking on a new important function. They begin to secrete the hormone progesterone, which acts on the lining of the uterus and prepares it for the possibility of implantation. A yellow pigment accumulates in the mass of transformed follicle cells, and they are now referred to collectively as the *corpus luteum* ('yellow body'). The corpus luteum enlarges for about ten days after ovulation as a result of stimulation by the pituitary, becoming some 2 cm in diameter, but if fertilization does not occur it begins to regress and involute. When the corpus luteum finally ceases to function it is converted into a mass of white fibrous tissue called the *corpus albicans*.

Administration of progesterone inhibits the process of ovulation, and this knowledge has been used in the preparation of oral contraceptives. The contraceptive 'pill' arrests the maturation of follicles and prevents ovulation, thus eliminating the risk of fertilization. On the other hand, ovulation can be stimulated by the administration of follicle-stimulating hormone followed by chorionic gonadotrophic hormone, and this treatment is sometimes used in women who are infertile due to a failure to ovulate.

Preparation of the uterus for implantation

The hormonal changes that occur during the ovarian cycle play an important part in preparing the uterus for the possible arrival of a developing conceptus. Oestrogen secreted by the Graafian follicles before ovulation causes the endometrium to proliferate. The first stage of this process involves repairing the damage caused during the previous episode of menstrual bleeding, and the epithelial cells of glands lying in the deeper part of the endometrium that still remains multiply and form a new epithelial covering to the endometrium. The endometrium now thickens, the glands enlarge, and at the same time the tissue becomes richly vascularized. After ovulation, progesterone (together with some oestrogen) produced by the corpus luteum stimulates the glands to secrete a fluid rich in glycogen, mucopolysaccharides, and lipids: all materials that are of value to a developing conceptus. The thickness and vascularity of the endometrium steadily increase during the two-week period that follows ovulation.

If fertilization occurs and a conceptus implants, the corpus luteum continues to secrete progesterone until the 4th month of pregnancy, but then it begins to shrink and becomes less productive. However, progesterone remains essential for the maintenance of pregnancy, and the placenta now takes over the role of secretion of progesterone previously fulfilled by the corpus luteum.

If, on the other hand, fertilization does not occur, the corpus luteum degenerates more rapidly, thus depriving the endometrium of the hormones required to maintain its highly vascular and secretory state. When this happens, the arteries supplying the superficial layer of the endometrium constrict for a time and the dependent tissues begin to deteriorate. The arteries then reopen, causing disruption of the deteriorating superficial tissues and the consequent shedding of cellular debris and blood into the lumen of the uterus. This produces the menstrual flow that generally occurs two weeks after ovulation without fertilization.

Chromosomal changes during gametogenesis

Cells of the human body usually contain forty-six chromosomes: the *diploid* complement. They can be subdivided into twenty-two pairs of *autosomes* plus one pair of *sex chromosomes*. The autosomal pairs differ in size when compared with each other, and they can be tabulated and numbered with regard to size, but the members of any given pair are equal in size. In the female the sex chromosomes — designated XX — are also equal in size, but in the male there is an X chromosome and a much smaller Y chromosome making up the pair of sex chromosomes. Thus the genetic complement of the female can be summarized as 44 + XX, and that of the male as 44 + XY (*see Figure 69*).

The arrangement of the chromosomes into pairs, so-called *homologous pairs*, is more than just a convenience for classification: it also relates to the origin of the chromosomal complement at fertilization when one member of each pair is provided by the maternal gamete and one by the paternal gamete.

There are two main types of cell division that occur in mammalian cells: *mitosis* and *meiosis*. In mitosis each daughter cell receives a diploid complement of chromosomes, but in meiosis each cell receives only *half* the usual number of chromosomes: a *haploid* complement. Meiosis is only seen during formation of the gametes.

As a prelude to mitotic cell division, copies are made of each of the chromosomes. (Details of this process of *replication* are given on p. 133.) The paired replicas are called *chromatids*, and during cell division they separate so that each daughter cell receives forty-six chromatids. The groups of chromatids become enclosed in nuclear membranes and are referred to once more as chromosomes.

The spermatogonia and oogonia possess forty-six chromosomes. They divide mitotically to form primary spermatocytes and primary oocytes, both of which also have forty-six chromosomes apiece. Before these cells divide to form secondary spermatocytes and oocytes, the chromosomes replicate in the usual way to produce pairs of chromatids. However, the chromatids remain combined in pairs instead of separating. As cell division begins, the chromatid pairs assemble in the middle of the cell and become arranged with their homologous counterparts. The *spindle apparatus* develops and moves the chromatid pairs to opposite poles of the cell, so that for each homologous group one pair of chromatids goes in one direction while the other pair goes in the opposite direction. Thus a secondary spermatocyte or secondary oocyte receives in effect only twenty-three chromosomes, even though each chromosome at this stage consists of a pair of identical chromatids. In the male, 22 + Y chromosomes go to one secondary spermatocyte, while 22 + X chromosomes go to the other. In the female, 22 + X chromosomes go to the secondary oocyte, while 22 + X chromosomes go to the relatively tiny first polar body. These events are referred to as the *first meiotic division*.

The *second meiotic division* occurs when the secondary spermatocytes divide to form spermatids and the secondary oocyte divides to form an ovum and a second polar body. (The first polar body may also undergo the second division to form two additional second polar bodies.) The members of the chromatid pairs simply separate during the second meiotic division — there is no further replication of DNA — and one member moves to one pole while the other member moves to the

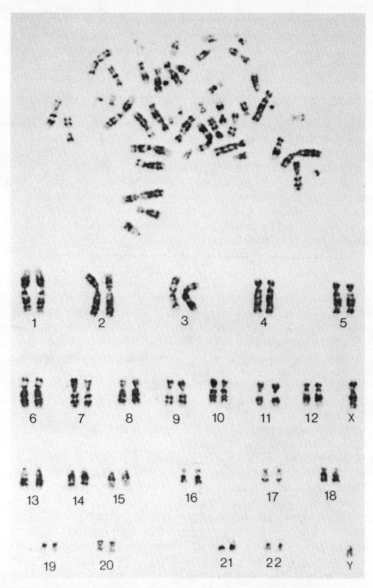

Figure 69. Chromosomes from a human cell undergoing mitosis. In this case the cell came from a male, and after sorting and matching of the chromosomes the normal 44 + XY complement is apparent. [From Sumner, A. T., et al. (1971) *Nature New Biology*, **232**, 31]

other pole of the dividing cell. Thus each daughter cell receives twenty-three chromatids.

As a result of this special sequence of cell divisions, a single spermatogonium gives rise to eight spermatozoa, of which four are 22 + Y and four are 22 + X. On the other hand, an oogonium eventually gives rise to two ova with 22 + X chromosomes and up to six polar bodies with 22 + X chromosomes — assuming, of course, that ovulation and not atresia occurs.

Fertilization of an ovum by a spermatozoon bearing a Y chromosome produces a male child, whereas fertilization by an X-bearing spermatozoon produces a female child.

Transport of the gametes to the site of fertilization

The sensory stimuli that arise during intercourse initiate reflex contraction of the muscle fibres in the vas deferens, the seminal vesicles, and the prostate, and the suspension of spermatozoa is passed into the first part of the urethra. The seminal fluid contains fructose and is further enriched by the secretions of several small glands. It is finally ejected from the penile urethra by rhythmic muscular contractions. The outlet sphincter of the bladder contracts during ejaculation to prevent a reflux of the fluid into the bladder.

The 3–4 ml of ejaculate contain 300–400 million spermatozoa, but of all those that start the race only one completes the course and fertilizes the ovum. The acidity of the vaginal secretions tends to reduce sperm motility, and it is therefore necessary for the spermatozoa to be transported rapidly to a more favourable environment. Under their own power spermatozoa can maintain a speed of about 2 mm per minute, but it is thought that at least part of their journey to the uterine tubes is aided by muscular contractions in the walls of the uterus and the uterine tubes, which might draw the spermatozoa along more rapidly. The environment within the uterus and tubes is more alkaline than that in the vagina and thus more favourable for the spermatozoa, but even so their life span after ejaculation is less than two days.

The significance of the astronomical number of spermatozoa present in the ejaculate is not clear, but it has been suggested that it may represent an evolutionary relic. In simple aquatic forms the mobile spermatozoa are released into the water, and their concentration becomes greatly diluted as they move away. Thus to give a reasonable probability of fertilization of the yolk-laden, relatively inert ova, it is necessary to release enormous numbers of spermatozoa. In the case of land animals fertilization occurs internally, and with a well-defined reproductive tract there is less risk of dilution and wastage. Another idea is that the large number of spermatozoa competing for fertilization provides an opportunity for 'selection of the fittest' for the important role of fertilization.

Immediately after ovulation the ovum is drawn into the open bell-shaped end of the uterine tube. The fimbria have been observed to clamp down closely over the ovary just before ovulation, and this action may increase the chances of the ovum entering the uterine tube rather than becoming lost in the peritoneal cavity. The ovum is transported towards the uterus by currents of fluid generated by the ciliated lining of the uterine tube, and it has usually traversed the first one-third of the tube before fertilization occurs. If fertilization does not occur within twenty-four hours of ovulation the ovum degenerates.

3 Abnormal Development

The Incidence, Causes, and Study of Abnormal Development

In view of the speed and complexity of early development it is hardly surprising that things can go wrong. Sometimes the abnormalities are so severe that the baby is stillborn or dies soon after birth; in other less severe cases the fault can be completely remedied by surgery or medication. Lying between these extremes there are abnormalities that are not severe enough to be fatal but cause great distress to the affected child and those responsible for caring for him.

Approaches to the study of abnormal development

Throughout historical times there has been a deep interest in abnormal development, and many ideas have been put forward to explain the phenomena. The Babylonians, and later the Greeks and Romans, looked upon the occurrence of malformed babies and domestic animals as omens for the future or punishments for the past, particularly if their distribution displayed a noticeable pattern. The Hebrews held that cohabitation with the devil was responsible — an outlook that would surely strengthen the feelings of guilt experienced by the unfortunate parents. Aristotle was prepared to consider the possibility that natural causes may be involved and, as mentioned in the introductory section, felt that prenatal development of a baby could be influenced by the experience of its mother during pregnancy. He would probably have supported the attitude expressed in the following extract, which comes from a scientific paper published in 1888. It relates to a newborn colt that possessed a single large eye placed in the centre of its forehead: 'As the mare has been kept near the railroad, it is believed by some that this strange freak of nature was caused by sudden fright at the train after night as the eye somewhat resembled the headlight of a locomotive.' Although shadows of these earlier ideas are still in evidence today, the original surrender to supernatural causes has been replaced by a scientific approach aimed at treatment, early detection, and prevention. However, in modified form the emphasis on the importance of the mother's environment remains.

The study of abnormal development is called *teratology*. This term is derived from the Greek *teras*, which means an object that arouses awe or wonder — a

112 Before Birth

portent. Information about abnormal development can be derived from a number of sources. In the case of human populations, considerable attention is being paid to the incidences of different malformations and their relationship with geographical areas or environmental factors. This statistical approach is known as *epidemiology* and has already revealed several significant patterns and correlations. Other clues have come from experiments carried out on the embryos of laboratory animals, and together with epidemiological information these observations are providing a clearer understanding of the causes of abnormal development.

Incidence of malformations

Before stating how many abnormal babies are born it is necessary to define how the distinction is made between abnormal and normal. One result of sexual reproduction is the occurrence of variation in a population; i.e. a new individual is slightly different from his parents and brothers and sisters. However, such variation is usually small and consists of slight differences on one side or the other of a mean value that is typical of the species. If we take a simple measure such as, for example, height, we find that the heights of most adult individuals come within a fairly narrow range and that the more extreme values are comparatively rare. When the results are plotted graphically, as illustrated in *Figure 70*, a distinctive type of distribution becomes apparent. This type of curve is known as a *normal distribution* and is commonly seen in sets of biological data. Since the population shows an unbroken curve of values we are faced by the problem of drawing arbitrary dividing lines if we want to establish a 'normal' range of heights, outside which somebody is 'abnormally' short or tall.

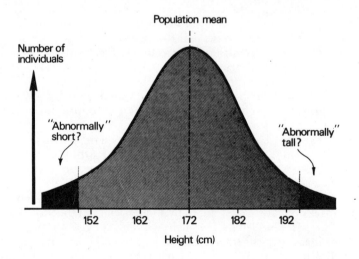

Figure 70. A normal distribution. The height of most adult males lies within a narrow range, say 152 cm to 192 cm, and fewer people are found with heights much above or below this range. (The figures used for this example are imaginary, not real.)

The same problem of arbitrariness applies to considerations of the incidence of developmental abnormalities, because usually there is a complete spectrum linking abnormal with normal variation. However, with this in mind, we can define what is meant by *congenital abnormalities* and to give some idea of their incidence.

A congenital abnormality is a structural or functional fault that originates before birth and seriously interferes with the subsequent everyday life of the affected child.

Using this criterion it has been established that about 2 per cent of babies are abnormal at birth. Of that number some soon die, but when the surviving children are assessed as they grow, more abnormalities become apparent. By the end of the first year, about 4 per cent of the infants are found to be abnormal, much of the increase being due to the detection of more subtle abnormalities, especially those of the special senses. It is impossible at this time to obtain more accurate figures, because different methods are used when recording the information and there is still disagreement about what is and what is not abnormal. These factors make it difficult to compare figures obtained in different regions. However, it seems that differences in incidence do indeed exist from region to region and from one racial group to another, and that the proportions of different abnormalities in the overall total are not the same in different populations. For example, although the overall incidence of abnormalities is similar in Birmingham and Japan, it has been noted that mongolism and malformations of the central nervous system occur more frequently in Birmingham than in Japan, while in Japan cleft lip, cleft palate, and congenital dislocation of the hip are more common.

Causes of abnormal development

It is now clear that abnormal development is the result of an interplay between genetic and environmental factors. Thus it is not only the genetic complement of the conceptus that must be considered, but also the environment in which it develops. The comparative significance of these factors differs from abnormality to abnormality. In some cases (e.g. mongolism) the genetic contribution is very large, while in other cases (e.g. abnormalities triggered by certain drugs) the environment is the major contributor. However, only about 25 per cent of all abnormalities are clear-cut in this way, and the majority seem to be a result of a more complex and subtle interaction: they are multifactorial. The following discussion is based largely on the more clearly understood abnormalities.

Genetic factors in abnormal development

A normal human somatic cell contains forty-six chromosomes. When straightforward techniques of chromosomal study became available it was soon noticed that a correlation exists between disorders of the chromosomes and certain characteristic abnormalities, e.g. mongolism and intersexuality. In some conditions there are too many or too few chromosomes, while in others there are faults in the structure of a single chromosome or a number of chromosomes.

Before considering the more common genetic syndromes it is worth noting that many conceptuses that have chromosomal aberrations are spontaneously aborted (miscarried). It has been estimated that at least 4 per cent of conceptuses show some form of chromosomal fault, but it seems that 90 per cent of those so affected are eliminated by spontaneous abortion. Thus, by the time of birth, the proportion of genetically abnormal babies is only 4:1 000. About 50 per cent of these are mongoloid and can be diagnosed immediately, but the remainder, especially those with abnormalities of the sex chromosomes, are usually less readily identified. The high incidence of genetic faults in spontaneously aborted conceptuses has lead to the proposal that a mother is able in some way to recognize and reject abnormal embryos, but the mechanism involved is still a mystery.

An additional chromosome: trisomy

Sometimes an additional chromosome is present in every cell. The effect will depend on which chromosome is overabundant. The most common defect of this type is the presence of an additional chromosome 21, a condition that is variously known as *trisomy 21*, *Down's syndrome*, or *mongolism*. In Europe the overall incidence of mongol births is about 1:600, but it is well known that the risk of a mother having a mongol child rises sharply as she reaches the end of her child-bearing period. For mothers aged 45 or more, the incidence is about 1:50, whereas during the middle reproductive years − say 20 to 35 − the risk is only 1:1 000. Because there is a strong link between incidence and maternal age (paternal age is not a factor) it is believed that the chromosomal fault stems from an aberrant division during oogenesis. It will be remembered that the oocytes shed from the ovary during the reproductive years were already present before birth. Thus an egg shed by a 45-year-old woman is quite an aged cell that has been lying dormant for many years. This may explain why chromosomal abnormalities become more frequent when such a cell is called upon to complete its meiotic divisions in readiness for fertilization.

Many of the differences between a mongol and a normal child are relative rather than absolute. The head is small and there is severe mental retardation, although the child usually responds positively to affection. The eyes tend to be small and slant upwards away from the nose. The mouth is also small, and the larger-than-normal tongue may protrude. The hands are short, broad, and marked by a deep transverse crease. More serious physical malformations associated with this syndrome include heart defects, umbilical herniation of the intestines, and poor development of the duodenum. An individual with Down's syndrome rarely lives beyond early adulthood.

Other autosomal trisomies include trisomy 17−18 and trisomy 13−15, but these are rare.

Trisomy of the sex chromosomes can also occur. In *Klinefelter's syndrome* the cells contain XXY chromosomes instead of XY. Initially the boy develops normally, but at puberty the testes remain small and the seminiferous tubules may show degeneration. Consequently, individuals with this condition are sterile. However, the interstitial cells of the testes − the cells that produce the male sex hormone − are normal, and male secondary sexual characteristics appear, although there may in addition be breast enlargement and stunting of growth.

Absence of a chromosome: Turner's syndrome

Occasionally a chromosome is absent. If there is a single X chromosome present, giving a sex chromosomal complement of XO, the individual shows the characteristics of Turner's syndrome. The low level of oestrogen associated with this condition results in the external genitalia, breasts, and ovaries remaining poorly developed after puberty. Other defects may also be present: skeletal deformities, heart abnormalities, mental retardation, and webbing of the skin of the neck.

Abnormal genes

Some genetic faults are more subtle than the chromosomal abnormalities outlined above and can be traced to a defect in a single gene. However, they can produce severe abnormalities ranging from skeletal defects (e.g. claw hand, dwarfism, cleft palate) to albinism and deaf-mutism. As may be expected, a harmful gene tends to be self-eradicating, since the affected individual has a poor chance of survival or is unable to reproduce, but if the gene is recessive it can remain in the population for long periods. This is because the defect is only expressed in those rare individuals in which both chromosomes of a homologous pair contain the gene. Occasionally it is necessary for certain environmental factors to be in existence in order to potentiate, or bring out, the genetic weakness.

Abnormalities due to environmental factors

Some abnormalities can be attributed to environmental factors, i.e. chemical or physical agents that are able to change a genetically normal conceptus into an abnormal baby. When considering human populations it is very difficult to decide which factors are having a harmful effect because of the complexity of the environment in which we live. However, some factors have been positively identified, and many others are suspected of being harmful.

Infection

The *rubella virus* is now known to be teratogenic. If the mother is infected with this virus, which is better known as german measles, there is a risk that her child may be born blind or deaf or in severe cases have malformations of the brain and heart. The abnormalities produced by this disease depend on the stage of pregnancy reached when the mother becomes infected. This link between rubella and a high risk of congenital abnormalities was first noted in 1941 by an Australian doctor called Gregg. Later studies indicated that the risk of the baby being malformed is 83 per cent when infection occurs during the 1st month of pregnancy, 61 per cent in the 4th month, and 10 per cent towards the end of pregnancy. To an adult rubella is a relatively mild infection with undramatic symptoms, but the high risk of prenatal damage has led to the development of a vaccine to protect pregnant women. A single infection with rubella usually (but not always) renders an individual immune to further attacks, and for this reason young girls are sometimes encouraged to come into contact with infected friends before they reach child-bearing age.

Other viruses that have been implicated are influenza, infective hepatitis, and mumps, but the picture in these instances is complicated and there are probably other environmental and genetic considerations. (Very recently it has been shown that influenza virus causes brain damage in monkeys infected during foetal development.)

Infections other than viruses can pass from the mother to the foetus. Maternal infection with the protozoan parasite *Toxoplasma gondii* is known to affect the growth rates and later development of the central nervous system and the eye, but it does not produce gross defects. Syphilis can also be transferred to the foetus via the placenta, usually after the 4th month of development. If this happens the growth of the foetus may be retarded, but, as with Toxoplasmosis, morphogenesis is usually completed successfully before the damage is caused and gross abnormalities do not arise.

Drug-induced abnormalities

The thalidomide tragedy of the early 1960s drew attention to the potential danger of man-made drugs. Thalidomide was used to prevent morning sickness in pregnant women during the first months of pregnancy. Some two years or so after its introduction it was noticed that the incidence of certain limb malformations had risen sharply, and it has since been estimated that 7 000 babies were affected. The malformations were of a type that is normally very rare, and this was a major factor in the discovery of the teratogenic action of thalidomide. Less conspicuous malformations might have failed to arouse attention so quickly. The drug had been tested by the standard methods then in use, but it was found later, after the realization that thalidomide is teratogenic in man, that embryos of different species respond very differently to maternal medication with a drug, and that the embryos used in the original tests were resistant to thalidomide. There has since been a revision of the way in which drugs are tested, or screened, before being used on the human population.

Thalidomide exerts its maximum effect during the first 3 months of pregnancy, causing shortening, stunting, or even complete absence of the limbs. In addition, the heart, intestines, and ears are sometimes affected.

Other drugs appear to be teratogenic under certain conditions. One of them — *aminopterin* — was used to induce abortion until it was found that if abortion was not produced there was a high risk of the foetus being grossly malformed. Therefore use of the drug as an abortificant has stopped. Other dangerous drugs include quinine, mercaptopurine, and cyclophosphamide, but their action is less clear-cut than that of thalidomide.

Hormonal factors

Hormonal factors can also disrupt normal development. Progesterone-like hormones that are given to mothers if they are about spontaneously to abort their foetus can produce masculinization of female babies. If a mother suffers from diabetes mellitus the risk of abortion or still-birth is high. This type of diabetes is a result of a low level of insulin in the blood, with the consequence that the glucose balance of the body is upset. Modern treatment can control the condition in the mother, but the risk for the foetus remains. Babies born to diabetic mothers are usually larger than normal and often show hyperplasia of the islets of Langerhans in the pancreas — the beta cells of the islets being the source of insulin. However, gross defects do not seem to be associated with maternal diabetes.

Irradiation

If an embryo or foetus is exposed to ionizing radiation, e.g. X-rays, there is a high risk that it will develop abnormally. The outcome depends largely on the dose received and the stage of development reached before exposure to the radiation.

When ionizing rays pass through a cell they impart sufficient energy to break up or activate molecules, especially large molecules such as proteins and nucleic acids. In this way irradiation may permanently damage chromosomal structure.

The damage may be great enough to lead to the immediate death of the cell, or it may result in a failure of the cell to divide normally at a later time. Smaller doses produce subtler changes and may result in abnormal behaviour of the cell rather than cause death. Dividing cells are particularly vulnerable to damage by ionizing rays, perhaps because the chromosomes have a less stable configuration during division, and thus it is not surprising that embryos — with their high proportion of dividing cells — are very sensitive to irradiation. On the other hand, highly differentiated cells are comparatively resistant.

Diagnostic irradiation uses a low dose of X-rays that is probably not harmful to a foetus, but radiological examinations of the mother are usually avoided if possible, especially early in pregnancy. However, the large doses of radiation needed for the treatment of cancer were soon found to produce malformations in embryos and foetuses, with the central nervous system, special senses, and skeleton being especially sensitive. Now that the dangers of irradiation are better understood great care is taken to ensure that a woman is not given radiotherapy if she is pregnant, and radiation-induced abnormalities have become very rare. The last recorded examples followed the atomic bomb explosions in Japan in 1945, when there was an increased incidence of disorders of the central nervous system in babies born to mothers who were exposed to radiation from the blast.

In addition to the immediate dangers of irradiation during development there is also a long term risk, since irradiation of the gonads, even with a very low dose, may gause gene mutations in the gametes or their precursors. Mutations can be beneficial as well as harmful, but there is always the risk that a mutation could give rise to congenital malformations in future generations.

Mechanical pressure

Throughout most of pregnancy the amniotic fluid provides a buoyant medium within which the foetus can move and grow quite freely. However, if the quantity of fluid is significantly less than normal, or if there is a large uterine tumour, the protruding parts of the foetus may suffer increased pressure, and this can cause abnormal moulding of the body, especially the limbs. If the limbs are pressed against the face of the foetus the facial features may also become very distorted, but often quite serious malformations of this type correct themselves spontaneously after birth as the child grows.

Maternal antibodies

The human body has defence mechanisms that enable it to recognize and often to destroy any foreign materials that enter it. The foetus, since it has a different genetic makeup from the mother, contains many recognizably different molecules and would be in danger of rejection by maternal defence mechanisms if it were not for the specialized nature of the placenta. The placental barrier prevents recognition of the foreign proteins by the mother. However, the barrier is not always totally effective, and sometimes the mother begins to synthesize antibodies against foetal materials. The antibodies tend to act against specific cell types, and certain organs of the foetus (e.g. the thyroid gland) may be selectively destroyed. If that occurs the baby will be irreversibly damaged by the time of birth.

In another condition, called *foetal haemolysis,* the foetal blood cells and the tissues that produce them are destroyed by maternal antibodies. This happens when blood cells escape from the foetus by passing through the placenta into the maternal bloodstream, where they may alert the mother's immune response system. Sensitization of the mother to foetal blood occurs most commonly when the mother has rhesus-negative blood and the foetus (and thus the father) is rhesus-positive. Specific antibodies are synthesized, and these return to the foetus quite readily. If a large proportion of the foetal blood is destroyed the baby is very anaemic at birth and its tissues suffer acutely from a lack of oxygen. A 'blue baby' of this type may require an immediate blood transfusion if it is to survive. Recently, techniques have been developed that make it possible to transfuse the foetus while it is still *in utero*. This procedure is valuable if a high level of specific antibody is found in the blood of a rhesus-negative woman, and it can reduce the extent of the damage that is caused by foetal haemolysis.

It is rare for a severe degree of haemolysis to occur in a first-born child. This is because there is insufficient time for a high antibody level to develop in the maternal bloodstream. In subsequent pregnancies the risk is much greater.

Not all the antibodies that pass from the mother to the foetus are harmful. Some of them confer a beneficial immunity to the foetus against diseases such as polio and measles. Other antibodies are transmitted in the mother's milk during the first weeks after birth. These help to protect the neonate during the period when its own defence mechanisms are being perfected.

Other factors linked with abnormal development

Statistical studies of human populations have shown correlations between congenital abnormalities and a number of environmental and genetic factors. The connections tend to be subtle and are best considered as interesting observations whose significance is not clearly understood at this time. For example, there are a number of maternal factors that seem to influence the incidence of malformed babies. The link between maternal age and mongolism has already been pointed out, but there is a more general relationship between especially young and old mothers and an above-average risk of bearing a malformed baby. In addition, first-born children are more likely to show defects of the central nervous system than second, third, or fourth children – thus parity is important in some way. As may be expected, the mother's diet seems to influence prenatal development, and there are indications that deficiencies of vitamin A or folic acid can cause abnormalities.

Racial and geographical differences in the incidence of some abnormalities have been noted, but at this stage it is difficult to assess the relative importance of genetic and environmental factors in the figures. Similarly, the high risk of recurrence of congenital malformation in a family that already has one affected child (e.g. a forty-fold increase in risk for cleft lip and cleft palate, a ten-fold increase for pyloric stenosis) may be due either to a genetic tendency or to prevailing adverse environmental conditions for the family.

There are even correlations between the weather and abnormalities. A severe winter tends to be followed by an upsurge in the incidence of anencephalus and spina bifida, while a good summer has been associated with an increase in pulmonary stenosis and abnormalities of the large intestine. It seems very unlikely that the weather is a direct cause of abnormal development, and it is more probable that it operates indirectly by influencing other teratogenic factors such as infection.

Experimental teratology

The human environment is already very complex and will probably become increasingly so as the number of man-made materials increases. In addition, the genetic makeup of a population is generally richly diverse, and it continuously receives new genes as families migrate and intermarry or as mutations occur. Thus, when studying the factors that cause congenital abnormalities it is necessary to look for patterns or correlations in observations derived from a bewilderingly complex interaction between heredity and environment. Only rarely can a specific factor be pinpointed, as occurred in the case of thalidomide.

Fortunately, clearer and less ambiguous answers can be obtained from less complex experimental situations set up in the laboratory. Genetic and environmental factors can be better controlled in such a situation, and, theoretically at least, a single variable can be selected and manipulated. Thus any change that occurs in the system can be directly attributed to the changed variable, although in practice it is still not easy to be sure that every variable except the one being studied is adequately controlled. Furthermore, it can be argued that these highly simplified situations bear little or no resemblance to the human situation, so that therefore the results are of little value. However, a number of valuable principles have emerged from these experimental studies.

Early in this century it was noted that embryos of fish, amphibia, and chicks developed abnormally when environmental conditions were manipulated. It was found that high temperatures, radiation, and various chemical treatments either were lethal for the embryos or caused malformations. In 1921 Stockard published a paper in which he put forward the concept of '*critical periods*', a concept that remains valid to this day. Using a variety of treatments he found that developing organs and systems are not equally susceptible to teratogens throughout their formation but pass through one or more sensitive periods. Most of these critical periods occur during the embryonic phase and usually coincide with bursts of cellular proliferation or rapid morphogenetic movements. When the morphogenesis of an organ is complete and differentiation is well advanced, the organ's resistance to teratogens increases. The different time of appearance of one structure compared with another and their differing rates of development result in a sequence of critical periods.

Real interest in experimental teratology arose in the 1940s, and it is now a flourishing branch of embryology. Many species and many teratogens have been tested. Obviously there are dangers in extrapolating from other species to man, but the techniques that have been established are undoubtedly useful, not only from the aspect of pure research but also in applied situations such as the screening of new drugs.

Generalizations that have emerged from experimental teratology

The response of an embryo to a teratogen depends on the following factors.

1. *The developmental stage when treated.* As the conceptus passes down the uterine tube towards the uterus it is quite resistant to teratogens. There has only been one report of abnormalities being initiated during the pre-implantation period, and that was in rats raised on a diet deficient in zinc. Usually the conceptus is able to implant and develop successfully even if the mother is given treatments that are teratogenic at later stages of development, provided that the dose is not large

enough to kill the conceptus outright. After implantation, however, and during the period of major organogenesis that follows, the embryo is very sensitive to teratogenic agents. When the critical periods have passed and differentiation is well under way, the susceptibility to teratogens becomes less once more.

2. *The teratogen used.* It was originally thought that all teratogens could produce equivalent effects if they were given at an appropriate dose level and stage of development. This view has been modified more recently as a result of work on mammalian embryos, and it now seems that most teratogens produce a characteristic pattern or syndrome of defects and are sometimes unable to produce all known defects. It is possible that these differences stem from individual characteristics of the extraembryonic membranes, especially the placenta, rather than indicate a difference in response by the embryonic tissues. There is, however, a wide degree of overlap in the action of teratogens, and even very different factors such as radiation and drugs can produce identical abnormalities. In this respect, the embryo resembles a watch: if the back of a watch is removed and particles of sand, sawdust, or iron filings are placed in the mechanism, it goes wrong, regardless of the precise nature of the material introduced.

3. *The dose administered.* The effect of a given teratogen depends greatly on the size of the dose administered. Usually there is a range of doses that can cause abnormalities, and outside this range the effect may be lethal, negligible, or even beneficial. Alternatively, there are environmental factors that become teratogenic if they exist in quantities that lie on either side of a particular range of values. An interesting example is given by vitamin A. It was found by experiment that vitamin A is essential for normal development, and that if it is deficient in the maternal diet the embryos develop abnormally or not at all. On the other hand, it is possible to have too much vitamin A in the diet, since hypervitaminosis A is also teratogenic.

4. *The dose rate.* The rate at which a particular quantity of a teratogen is administered will also influence the response of the embryo. If the dose is given in a number of fractions the result is less severe than that obtained when the whole dose is given at once. This softening of the impact is partly due to repair processes that occur between the fractions, but a dose given in a number of milder fractions will affect a longer section of the sequence of critical periods, and this also will contribute to the observed differences.

5. *The species of embryo treated.* Not all embryos respond in the same way to a given teratogen. Once again it is possible that the differing properties of the extraembryonic membranes in different species account for these discrepancies.

Experimental teratology has provided many clues for those searching for teratogens in the human environment, and it has also helped to test hypotheses that have arisen as a result of studies of population statistics. To understand this relationship between epidemiology and experimental teratology more clearly, it will be advantageous to focus on a single malformation and study it in more depth. Spina bifida has been selected for this purpose.

The Problem of Spina Bifida

Spina bifida and its treatment

Spina bifida is a congenital malformation of the vertebral column with, in many cases, involvement of the spinal cord. It is now possible to treat an affected baby, but considerable controversy has arisen as to the advantages or disadvantages of doing so. Early operative closure of the defective region of the back can greatly increase the baby's chance of survival, but it has become clear that the quality of life that results is generally poor — neither fair to the child nor good enough to justify the extensive medical and educational care that must follow. It has been suggested that some form of selection is required, to ensure that operations are carried out only in those cases that will benefit most. However, a proposal of this sort raises the difficult ethical problem of where to draw the dividing line, and it also implies that doctors will be called upon to make the decision as to whether or not an operation should be carried out. At present it is the parents who have the right to decide, although obviously their decision is very dependent on the attitudes of the doctor who informs them of the situation during the first hours after birth.

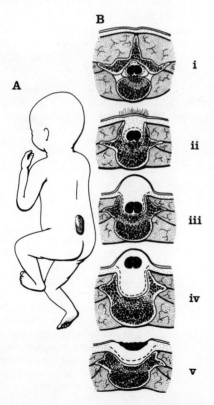

Figure 71. The external appearance and types of spina bifida. (A) A baby with spina bifida in the lumbosacral region. (B) Degrees of malformation. (i) A section through the lumbar region of a normal child. The spinal cord is shown in black and is surrounded by a sac of dura mater (dashed line). The spinal cord is contained completely within the vertebral canal, and the muscles of the back are evenly distributed around the bony vertebral column. (ii) Spina bifida occulta. There is a deficiency in the neural arch of one or several vertebrae, but the spinal cord and meninges are unaffected. (iii, iv, v) Spina bifida cystica. Three types are shown in order of increasing severity. In the most severe form the neural tissue is exposed as a flat degenerate plaque on the surface of the back. The vertebrae form a shallow gutter of bone beneath the plaque, and the extensor muscles of the back are displaced sideways and forwards.

122 Before Birth

The clinical significance of a particular case of spina bifida depends on the position of the spinal lesion and the degree of involvement of the spinal cord (*see Figure 71*). Usually the malformed area is in the small of the back or just below (the lumbar or lumbosacral region), but any part of the spine may be affected. In general, the nearer the lesion is to the head, the more serious are the consequences. Sometimes only the bony part of the spine is malformed, and the spinal cord is unaffected. There are rarely clinical complications with this type of spina bifida — *spina bifida occulta* — and many people have it without being incapacitated in any way. When the membranous coverings of the spinal cord or the cord itself is malformed — *spina bifida cystica* — the outlook is less hopeful, and the greater the disorganization of the spinal cord, the poorer the outlook becomes.

If an operation to cover the cystic area with skin is not carried out, most babies with spina bifida cystica die before reaching 3 years of age. The cause of death is usually infection of the central nervous system via the malformed vulnerable region of the spine. Studies indicate that only 3–15 per cent will survive, and even then the majority of the survivors are confined to wheelchairs or completely helpless.

Postnatal surgery

Several early attempts were made to treat the condition. In 1887 Dr Morton reported some success with injections of a solution of iodine in glycerine into the spinal swelling. The result was obliteration of the fluid-filled blister by fibrous tissue. However, the initial improvement in survival was offset by problems of incontinence, paralysis of the legs, and swelling of the brain due to the accumulation of fluid within it. This latter condition, a form of *hydrocephalus,* is the result of a malformation of the hindbrain. Under normal circumstances clear cerebrospinal fluid is formed within the brain and then passes out through perforations in the roof of the hindbrain into a spongy layer that covers the brain. The fluid is continuously absorbed from this layer into the bloodstream and thus does not stagnate or accumulate. In hydrocephalus associated with spina bifida the fluid cannot escape from the hindbrain and begins to accumulate within the

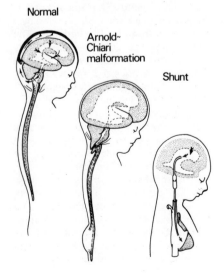

Figure 72. Control of hydrocephalus. *Left:* The normal flow of cerebrospinal fluid. *Centre:* The effect of the Arnold-Chiari malformation. Cerebrospinal fluid can no longer escape from the fourth ventricle because of the herniation of the hindbrain into the cervical spinal canal, and hydrocephalus develops. *Right:* The method of correcting hydrocephalus with a one-way shunt. A tube links a brain ventricle with the right atrium of the heart, and a one-way valve is placed subcutaneously behind the ear.

brain, causing it to enlarge. This increased pressure and consequent enlargement can rapidly cause damage, especially to the outer layer of grey matter, the cerebral cortex, which becomes stretched and thinned.

Another early method of treating spina bifida was to remove surgically the tumour-like swelling, but this approach was soon abandoned when it was realized that some of the excised tissue had still been functional and would have been better left in place.

A successful surgical approach did not become possible until techniques were developed to overcome the secondary complications associated with the spinal defect. In the early 1950s a method of controlling hydrocephalus was devised. It consisted of draining the excess fluid from the brain into the bloodstream or the abdominal cavity, as shown in *Figure 72*. An essential element of the shunt is a miniature one-way valve that prevents a return of materials back towards the brain, and the valve is engineered to open only when the pressure of fluid within the brain rises above a critical level. Most commonly, the shunt is arranged to pass out from the brain and skull and down below the skin into the neck, where it enters one of the large veins returning blood to the heart. The integral one-way valve is usually positioned behind the ear.

The success of this method of controlling hydrocephalus revived surgical interest in spina bifida. It was found that if the damaged flattened region of the spinal cord was sewn up into a tubular form and the skin of the back closed across the defective region, the risk of meningitis was greatly reduced and the chances of survival were correspondingly increased. In one study of operated children, it was shown that 89 per cent of those without hydrocephalus and 35 per cent of those with hydrocephalus survived until at least 7 years of age. Many of the children are now teenagers.

However, closure of the spinal defect and insertion of a shunt are only the first steps in a long series of operations. Other disorders are frequently associated with spina bifida, and they too require treatment. For example, where there is severe deformity of the spine and legs it is necessary to carry out major orthopaedic surgery if the child is to have any hope of walking. Also, it is quite common for kidney damage to occur. This is generally a consequence of poor or nonexistent innervation of the bladder, and the resultant urine retention with frequent episodes of infection make it necessary to divert the ureters to the abdominal wall where the urine can be collected directly. Thus, although children born with spina bifida cystica have today a good chance of survival, they face a sequence of major operations − on average about one every year. Complete cure is impossible, since the abnormal spinal cord has been degenerating for several months before birth and there is no way of replacing damaged nervous tissue. Thus postnatal treatment is at best palliative. In an attempt to avoid these problems, two alternative approaches are being investigated.

Prenatal detection and abortion
Early prenatal detection of abnormality may be followed by abortion of severely malformed foetuses if the parents consent. Several techniques of detection are being assessed and perfected, but two in particular are producing dramatic successes. *Amniocentesis* involves the extraction of a small volume of amniotic fluid, care being taken first to localize the placenta to avoid passing the syringe needle through it. This procedure can usually be carried out safely at any time after the

3rd month of pregnancy. By culturing cells from the fluid it is possible to detect chromosomal abnormalities such as mongolism, but in addition it is now possible to detect spina bifida or anencephalus early in pregnancy with a fair degree of confidence. The test depends on the amount of a protein called *alpha-foetoprotein* that is present in the fluid. This protein is present in much greater than normal quantities in the amniotic fluid surrounding a baby with spina bifida or anencephalus, and on several occasions the test has successfully detected a malformed foetus. However, this is a new and relatively untried method, and at present it is usual to check a positive diagnosis of abnormality with the second method of detection that is available — *ultrasound scanning* — before making an abortion. Ultrasound scanning uses the principle of an echosounder to build up an image of the shape and position of the foetus. The high-frequency sound waves are differentially absorbed or reflected by the tissues of the foetus and the mother and by the amniotic fluid, and the pattern is displayed as an image on a television screen. In skilled hands this technique provides as much information about the foetus as a normal X-ray film, but apparently with far less risk of damaging the developing tissues. It can accurately detect anencephalus and more severe degrees of spina bifida quite early in pregnancy.

Clearly, the ethical problems linked with prenatal detection and the abortion of selected cases are similar to those associated with postnatal selection of babies for operation. However, the foetus seems to be regarded by many people as less sacred than a newborn child, and such a proposition may therefore be more practicable.

Prevention

Before the prevention of spina bifida can become possible, it is necessary to detect the underlying causes and then attempt to remove or eradicate such factors. So far there is no clear understanding of the aetiology of spina bifida, but there are a number of promising clues that suggest that one day it will be possible to limit the number of babies that are affected. Evidence has come from studies both in man and in other animals.

Aetiology of spina bifida in man

When it was finally realized that genetic mechanisms alone were not to blame and that environmental factors can also have a significant effect on prenatal development, attention was focused on the relative importance of genetic and environmental factors in precipitating spina bifida.

The possibility of a genetic disposition to spina bifida is suggested by the racial differences in incidence. In a study of the Japanese population it was found that the incidence of spina bifida in babies was only 1:5 000, whereas in South Wales the incidence was twenty times as high (1:250). Observations of this type are complicated by the different environmental backgrounds of the populations studied, and it is not possible to make conclusions on these figures alone. It may be helpful to study families who migrate to new environments to see whether they continue to show the original incidence of spina bifida or whether the incidence changes towards that of the indigenous population. Preliminary findings suggest that the incidence of spina bifida in American Negros and in West Indian immigrants to Britain is closer to the generally low incidence characteristic of

West African populations, from which they were originally derived, than to the higher incidence in the indigenous Caucasian populations. However, there is evidence that the differences are diminishing. Once again it is necessary to be cautious in interpretation, because although some environmental factors (e.g. climate) are changed immediately for the migrant family, others (e.g. diet) may change much more slowly, and thus it can still be argued that the environment is not completely the same for the two groups being compared, even though they live in the same region.

A genetic tendency to spina bifida is also implied by the increased risk of a second child with spina bifida being born into a family in which there is already one affected child. This risk of recurrence of spina bifida is six to seven times greater than the incidence in the general population, and if a family has two affected children the risk then soars to a 1:10 chance that the next baby will be affected. Thus there seems to be a familial concentration of spina bifida, although it is uncertain how much of this is due to a genetic bias and how much is the result of a continuing adverse environment for the family.

It might be expected that if one member of a pair of identical twins is malformed, the other will show a comparable malformation, since both will have identical genes and both have developed in the same uterus. Surprisingly this is not so: on the rare occasions that spina bifida has occurred in identical twins, almost invariably only one of the twins has been affected. This puzzling lack of concordance in identical twins has not been satisfactorily explained, but it suggests that different conditions exist even within the same uterus: perhaps the maternal blood supply to one implantation site is poorer than to the other, or perhaps the presence of two rapidly growing embryos can result in competition for space, with one of the embryos fairing worse than the other and developing abnormally.

Therefore it is difficult to evaluate the evidence suggesting a genetic basis for spina bifida because of a number of complicating factors. Similarly, when trying to identify adverse environmental factors it is difficult to obtain a clear unambiguous pattern from the many observations that have been made. Several environmental causes of spina bifida have been proposed. For example, it was noted that an upsurge in incidence followed an epidemic of influenza in Ireland on one occasion, but a similar correlation has not been found on other occasions. Attention is now becoming focused largely on dietary factors. In 1972 a statistical link was demonstrated between certain food additives — magnesium salts in tinned peas and nitrites in cured meats — and the incidence of malformations of the central nervous system. In the same year a cogent hypothesis was proposed linking a disease of potatoes with spina bifida and anencephalus. Renwick had discovered a remarkable similarity between the distribution of *potato blight* and the incidence of these malformations. He postulated that the teratogenic substance involved is an antifungal compound, called solanidine, that is produced by potatoes in response to infection by blight or following injury to the tuber. Solanidine is known to be toxic to humans when taken in large quantities, and it has also been shown to be lethal to chick and rat embryos. More interestingly, an experiment stimulated by Renwick's hypothesis demonstrated that the incidence of cranial malformations was increased in the offspring of marmosets fed on a diet containing blighted potatoes. More recently however, contradictory evidence has accumulated and the initial interest in the potato blight hypothesis has waned.

The embryonic development of spina bifida

A limited amount of information about the embryonic development of spina bifida has been gained from studies of human embryos and foetuses. This material is often poorly preserved, and good examples of early embryonic stages are particularly rare, but two tentative hypotheses, illustrated in *Figure 73*, have been put forward to explain the primary lesion:

1. There is an opening of the already closed neural tube as a result of an increase in pressure of the fluid within its lumen, or
2. There is a failure of part of the neural plate to close and form a tube.

The second alternative has gained the greater support and has been borne out by observations in other species. The lack of suitably preserved human material makes it necessary to rely heavily on information derived from other species, although care must be taken when extrapolating from, say, rat embryos to human embryos. However, the similarity of different embryos, especially at the stage of neurulation, suggests that similar developmental mechanisms are at work (*see Figure 74*). It is only after major organogenesis has been completed that embryos begin to diverge rapidly along different species-specific pathways.

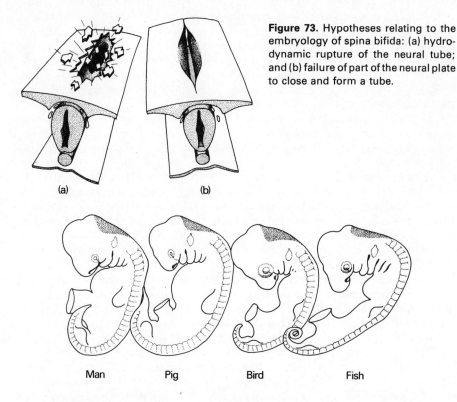

Figure 73. Hypotheses relating to the embryology of spina bifida: (a) hydrodynamic rupture of the neural tube; and (b) failure of part of the neural plate to close and form a tube.

Figure 74. A comparison of embryos from different species. Note their similarity during the early stages of organogenesis.

Spina bifida in other animals

As in the human context, both genetic and environmental factors are relevant in the aetiology of spina bifida in other animals. A familiar example of a spina bifida-like malformation in a domestic animal occurs in the Manx cat. In this strain the tail fails to develop, the pelvis is often abnormally proportioned, and the hindlimbs usually show some paralysis so that the cat moves by hopping like a rabbit. These features are much prized by the breeders and owners of Manx cats, which illustrates how opinions can differ as to what is desirable and what is abnormal.

There are other species with strains that show a marked tendency towards spina bifida, and those of the mouse and rabbit have been helpful in the study of the mechanisms involved. They have provided interesting information about quite early stages of abnormal development and have reinforced the view that nonclosure of the neural plate is the primary fault. Some clues were also obtained as to the sequence of development of the secondary faults such as malformations of the vertebral column and the hindbrain.

The most fruitful approach has been that of experimental teratology, in which embryos are exposed to harmful environmental factors under carefully controlled conditions and the ensuing abnormal development is analysed.

Experimentally-induced spina bifida

Many chemical and physical agents have been found to cause spina bifida in experiments, but two in particular have been intensively studied: trypan blue and ionizing radiation.

Trypan blue is a diazo dye that readily produces spina bifida experimentally, but it is unlikely to be a cause of spina bifida in the human context since it is not typically found in the general environment. However, it has revealed several interesting processes that may be of relevance to an understanding of how spina bifida develops. It appears to act by disrupting embryonic nutrition just before, or during, the embryonic period. The spinal lesions were generally localized in the lumbar or lumbosacral region as in man, and the general pattern of the primary and secondary defects was also comparable. The sequence of early changes that was observed will be incorporated into the generalizations given later.

Spina bifida was produced by *irradiation* in many experiments. The ease and accuracy with which X-rays can be administered has enabled precise information to be gained about the very early changes that occur. As in the case of trypan blue, irradiation is unlikely to account for many human malformations, but the insights it has given into the mechanisms involved are of great value.

Several other techniques have produced spina bifida. Folic acid deficiency (and folic acid antagonists such as aminopterin), hypervitaminosis A, nitrogen mustard, urethane, hydroxyurea, influenza A virus, RNA inhibitors, and substances that disrupt microtubules and microfilaments have all been found to produce spina bifida. As a result of all these experiments it is now possible to understand what happens during the few critical hours when a potentially normal conceptus makes the transition towards becoming an embryo with spina bifida.

The sequence of changes leading to spina bifida

The first visible sign of abnormal development is the failure of part of the neural plate to close. Since the notochord is usually normal and Hensen's node continues to migrate caudally away from the affected region, it would appear that things begin to go wrong soon after formation of that particular region of the neural plate. With regard to the possible causes of nonclosure, it has been shown that any physical or chemical factor that reduces the number of viable cells in the neural plate, interferes with microtubule or microfilament activity, or disrupts protein synthesis will tend to cause spina bifida. In the human context, spina bifida appears to be the outcome of a complex interaction of a number of factors; i.e. it is *multifactorial*.

After nonclosure a number of secondary changes occurs, illustrated in *Figure 75*. The sequence is as follows.

1. The affected region of the neural plate becomes thickened. This 'overgrowth' is partly due to an increase in cell division, but it is also the result of the enlargement of intercellular spaces and the steady accumulation of dead cells. The neural cells that remain viable attempt to establish the type of organization normally present in the neural tube, with the cells radially orientated around a central fluid-filled cavity. Frequently, a number of cavities are formed within the overgrown region and the neural cells form groups or 'rosettes' around them. This may explain the origin of a phenomenon that is sometimes associated with spina bifida: duplication of the spinal cord above or below the spinal lesion.

2. The wedge of neural crest cells that in normal embryos helps to close the roof of the neural tube (*see* p. 35) tends to hypertrophy in abnormal embryos at the cranial and caudal limits of the malformation. In some cases this change may partially compensate for the failure of the neural folds to close completely and bridge the abnormally wide gap between them.

3. The abnormal neural tissue separates from the adjacent ectoderm with which it is initially continuous, in much the same way that it would if a neural tube had formed successfully. However, an abnormal layer of fibrous material accumulates in the intervening zone, and this may help to strengthen or stabilize the malformed region, preventing extensive separation of nearby tissues and consequent damage.

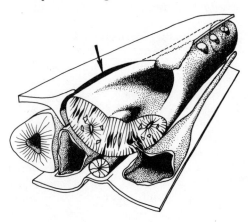

Figure 75. Early secondary changes in spina bifida. The exposed neural tissue has become thickened and shows regions of degeneration (black areas) and reorganization around secondary lumina. Enlarged blood vessels lie on each side of the malformed region, and one is shown invading the plaque. A fibrous interlayer is developed between the neural tissue and the adjacent ectoderm (arrow). Cells of the neural crest that normally migrate away from the line of fusion of the neural folds remain at the cranial and caudal limits of the lesion as hypertrophied wedge-shaped structures.

4. Within a short time of nonclosure the exposed neural tissue begins to degenerate. Large fragments of the tissue may be shed, and this process of rejection is occasionally aided by the growth of tongues of ectoderm inwards towards the midline just ventral to the neural tissue. The structures that would normally be innervated by the affected region of the spinal cord (e.g. muscles of the leg and the bladder) have to develop without a nerve supply or with only partial innervation, and in consequence they too develop abnormally.

5. The blood vessels developing in the mesoderm close to the neural plate become greatly enlarged, apparently in response to the accumulation of dead and dying cells, and they begin to penetrate particularly degenerate regions. This may indicate an attempt to remove the products of degeneration before they become toxic to the embryo.

6. The somites that lie alongside the abnormal region are less regular than normal. When the cells of the sclerotomal component migrate towards the malformed neural tissue, they are unable to establish vertebrae in the normal way. Since there can be no continuity across the dorsal aspect of the neural tissue as would be the case if a tube had formed, the bony arches that normally complete the vertebral canal fail to develop. The result is a gutter-shaped plate of bone in the malformed region. This configuration causes a redistribution of the powerful muscles of the back, which are normally arranged in a balanced way around the vertebral column. The muscles are forced to lie lateral or ventral to the bony plate, and as a result the spine, which is already weakened by its abnormal structure, is liable to bend and eventually collapse in the severest cases as a result of the unbalanced muscular stresses.

7. The tissues that under normal circumstances would have formed the wrappings of the spinal cord often become dilated by accumulations of cerebrospinal fluid, and this gives rise to the characteristic blister that is seen in the malformed region at birth.

8. The abnormality of the hindbrain that is often associated with spina bifida is less readily explained, but it is possibly related to the differing growth rates of the spinal cord and the vertebral column. During the foetal period the spinal cord grows less rapidly than the bony canal that surrounds it. In a normal foetus this differential growth can be compensated for quite readily by a gradual shift of the relatively free distal end of the spinal cord in a cranial direction. If the distal end of the cord is malformed, however, as in the case of spina bifida, the abnormal neural tissue remains anchored to the surrounding tissues and the spinal cord can no longer adjust its position from that end. Thus it has been proposed that a steady pull is exerted by the spinal cord on the hindbrain, with the result that the brain tissue begins to move back into the vertebral canal. Being wedge-shaped, the hindbrain would eventually block the canal, and the roof of the fourth ventricle, through which cerebrospinal fluid normally escapes from the brain, would become obstructed. As the pressure of fluid within the brain rises, it might be expected that there would be a further force acting on the herniated hindbrain − i.e. a push from above in addition to the pull from below − and that this would tend to increase the extent of the herniation. This hypothesis is at present the best explanation of the commonly observed association between spina bifida and hydrocephalus.

In conclusion

It can be seen that the controversy that surrounds the immediate problem of spina bifida is beginning to show the first hopeful signs of resolution. While many people feel that the approach of total operative care can produce more problems than it solves, sufficient experience has been gained as a result of operative programmes to make it possible to identify which children will benefit most from intensive care. In addition, it has been demonstrated that the defect can be detected early in pregnancy, and there are indications that methods of routine prenatal screening can be derived from these procedures. Progress has also been made towards prevention, which remains the ultimate aim, and there is a clearer understanding of the predisposing factors and mechanisms involved.

This rather detailed look at spina bifida is intended to illustrate the methods that are generally used in the study of congenital abnormalities, and it emphasizes the interdependence of the different approaches that are available.

Theoretical Aspects

Understanding Development

Different approaches

In the preceding description of development, occasional reference has been made to experimental procedures and results. The experimental approach has already provided a wealth of useful information about development and no doubt will continue to do so, hopefully at an ever-increasing rate as techniques more appropriate to the fragility of embryos are devised. However, the jigsaw puzzle is far from complete, and this may be a good time to consider which direction future studies could take.

Let us consider first the approaches that are currently favoured. In general terms, embryological experiments seem to fall into three main categories:

1. *The study of 'developmental mechanics'*, in which the physical properties of tissues and their interrelationships are considered. Attention is focused on morphogenetic processes — e.g. the mechanism underlying the folding of cell layers, the role of intracellular organelles such as microtubules and microfilaments, and the changes in cell-to-cell adhesiveness that seem to occur during development.

2. *The search for 'organizers' and 'inducers'*. The controlling influence that some regions appear to exert over adjacent regions during the embryonic period has prompted numerous attempts to isolate chemical mediators. This initially exciting approach has so far failed to give the major advances once expected, and although active compounds thought to be involved in the development of regional differences have been extracted, their chemical identities remain a mystery. In time it may be necessary to conclude that induction is effected not by specific information-carrying molecules, but rather by regional variations in the concentration of smaller molecules such as metabolites and nutrients.

3. *The analysis of abnormal development*. By observing what can go wrong during development a better appreciation can be gained of normal processes, particularly if the factor precipitating abnormal development and the stage at which it acts are known. Thus teratology provides a double-edged sword that can be used in analysing both normal and abnormal development.

Teratology is a relatively recent offshoot from the mainstem of experimental embryology and shows considerable promise, but there is another area of

132 Before Birth

research that has had even greater impact on our way of thinking, not only about development but also about biological systems in general. Holding the centre of this arena are two molecules known as *DNA* and *RNA*, and it is worthwhile now to consider their remarkable properties in some detail.

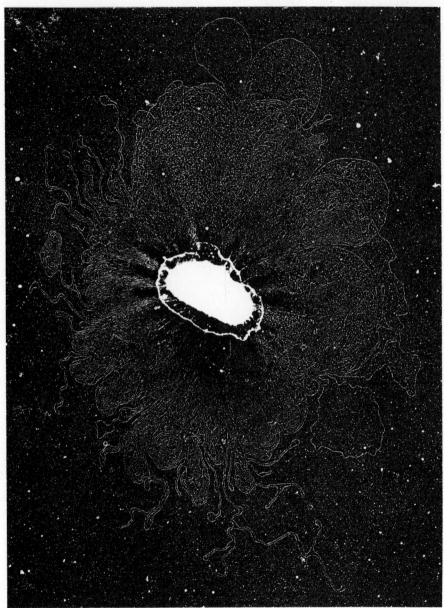

Figure 76. A bacterium that has been ruptured by osmotic shock to reveal its long chromosome. [From MacHattie, L. A., *et al.*, (1965) *Journal of Molecular Biology*, **11**, 648]

Deoxyribonucleic acid (DNA)

Chromosomes, DNA, and genetic information

The nucleus of a cell contains a number of structures called chromosomes. In human cells there are usually forty-six chromosomes, and they can be seen most readily just before a cell begins to divide. For a long time it has been realized that the chromosomes, or more specifically the molecules of DNA that they contain, are involved in heredity. Evidence of this came from a number of sources. For example, it has been demonstrated that foreign DNA inserted into a bacterium can cause a radical transformation of its structural and functional organization. It has also been noted that the quantity of DNA in the somatic cells of higher organisms is constant for a given species, even though the cells come from very different tissues, while a comparison of one species with another shows a characteristic difference in DNA content. In keeping with its function, DNA is a remarkably stable compound — studies with radioactive tracers have shown this — and it can therefore act effectively as a long term store for genetic information.

A chromosome is built up from two materials: a long molecule of DNA, and protein molecules known as *histones* (see *Figure 76*). However, the exact arrangement of these constituents is still being debated. Until 1973 it was believed that the thread of DNA formed the core of the chromosome with the histones forming a protective sheath around it. The composite strand was then believed to be tightly coiled. More recently, however, a new picture of chromosomal structure has emerged in which the histones are thought to be arranged like a string of beads with the DNA molecule wrapped tightly around them. The two interpretations of chromosomal structure are illustrated in *Figure 77*.

There is now a great deal of evidence that DNA carries species-specific information in a linear form along its length, and that this sequence can be subdivided into theoretical 'units of inheritance' known as *genes* or *cistrons*.

The structure of DNA

In 1953 Watson and Crick showed that DNA has a double helical structure. The breakthrough came when they were able to build a three-dimensional model that explained their X-ray diffraction patterns. (A fascinating personal account of the events leading up to the discovery is given by Watson in his book *The Double Helix*. It is also worth looking at Salvador Dali's painting 'Galacidalaciddes-oxiribonucleic'!) They showed that the helical sugar/phosphate backbones of the DNA molecule are linked by transverse bridges of paired bases: *adenine—thymine* and *guanine—cytosine*, as shown in *Figure 78*. The molecule thus resembles a spiral staircase with the base pairs forming the horizontal steps. This selective pairing of nucleotide bases is a result of their differing sizes and of the spatial arrangement of the hydrogen bonds that must form to ensure a stable linkage. Thus the sequence of bases attached to one sugar/phosphate backbone is complemented by a related pattern of bases on the adjacent spiral, the pattern being determined by the selective affinity of adenine for thymine and guanine for cytosine. Watson and Crick recognized immediately that a mechanism exists here whereby copying, or *replication*, of genetic information can occur.

Replication of DNA

Replication begins when the strands of the helix separate, exposing the two complementary sequences of bases, as shown in *Figure 79*. Free nucleotides

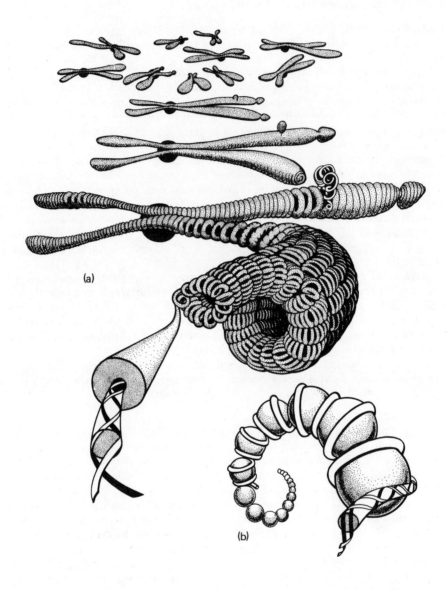

Figure 77. Two interpretations of chromosomal structure. (A) At metaphase, each chromosome consists of a pair of chromatids. Each chromatid consists of a protein-coated strand of DNA showing three orders of coiling. The protein sheath is composed of histones. [After Gray, H., (1973) *Anatomy*, 35th edn (Edinburgh: Churchill Livingstone)] (B) In a more recent model, there is a highly organized core of histone molecules around which the DNA strand is coiled. This chromatin 'string of beads' is believed to be further supercoiled in order to pack the enormously long DNA molecules within the confines of the nucleus. [After Lewin, R. (1975) *New Scientist*, **66**, 308]

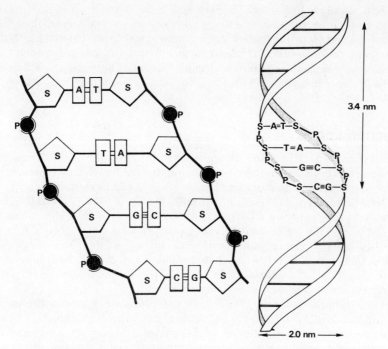

Figure 78. The Watson and Crick model of DNA. *Right:* A general view of the double helix. *Left:* A more detailed diagram of its organization. (A = adenine, T = thymine, G = guanine, C = cytosine, S = sugar, and P = phosphate.)

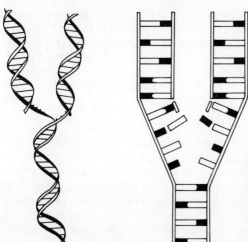

Figure 79. Replication of DNA. *Left:* The parent molecule unwinds, and new strands are formed using the exposed bases as a template. *Right:* A simplified diagram; the helical shape of the molecule has been ironed out to show that two identical molecules of DNA are formed by this process of specific-base pairing.

present within the nucleus attach themselves to the exposed bases in the appropriate way (adenine−thymine, guanine−cytosine) and a new sugar/phosphate strand is arranged along the free ends of the newly established base pairs. In this way two identical molecules of DNA are formed where previously there was only one. Each contains half the parent molecule, which has acted as a template for assembly of the remainder. Thus the information contained in the original molecule has been copied without loss.

The genetic code

The series of bases in DNA can be thought of as a linear sequence of information. This *genetic code* is built up from a four-letter alphabet, with each base representing a letter. Since the DNA in a mammalian cell is thought to contain some 3 000 million base pairs, there is potentially an immense store of information. However, allowances must be made for different types of *redundancy* of information. Clearly, not all the genes (nor indeed all the chromosomes) in a cell express the information they carry. Some remain dormant as the cell differentiates or are suppressed by the action of other, more dominant genes carried by homologous chromosomes. Also, the genetic code requires punctuation between adjacent genes, or sequences of genes, to prevent ambiguity.

It is now established that the genetic code provides instructions for the assembly of amino acids into proteins. The sequence of amino acids in a protein dictates its properties and role within the cell, and the types and proportions of proteins within a cell similarly impart a distinctive form to the cell − functionally as well as structurally, since most enzymes are proteins. Thus, in suggesting that DNA controls protein formation we are implying that DNA also controls cellular differentiation.

Current evidence suggests that an amino acid is specified by a sequence of three bases called a *triplet*. Since there are four distinct bases in the genetic alphabet, there are theoretically sixty-four different triplets available for use in the code, but it has been found that only sixty-one actually specify amino acids. Since there are twenty biologically useful amino acids, this indicates that there is sometimes more than one code word, or *codon*, for a given amino acid. This existence of alternative codons is sometimes referred to as 'degeneracy' of the code, because it represents another form of redundancy. The remaining three triplets do not code for amino acids but punctuate the information sequence, an illustration of which process is given in *Figure 80*. In the absence of punctuation the message could be rendered meaningless or misleading, as the following verbal analogy illustrates:

TIMEFLIESYOUCANNOTTHEYPASSTOOQUICKLY
TIME FLIES? YOU CANNOT. THEY PASS TOO QUICKLY!

Although genetic, or hereditary, information is usually carried by DNA, there are exceptions to this rule. For example, some viruses contain a nucleic acid of a different type called RNA. Poliovirus is an RNA virus whose particles have a molecular weight of only 8 million − it is thought to be the smallest self-replicating unit in nature − but in spite of its diminutive size poliovirus produces significant alterations in the organization of the cell it infects, converting it into a factory for producing new virus particles. This indicates that even a very limited amount of information can produce dramatic cellular changes. However, in most

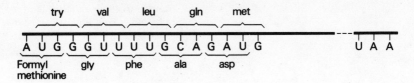

Figure 80. Punctuation of the genetic code. With a linear sequence of information it is important to know where to begin and where to finish reading. The triplet AUG is known as an 'initiation' triplet and specifies the start of a sequence of amino acids. At the other end there is a nonsense triplet UAA, which does not specify an amino acid. This acts as a full stop to the sequence. Without punctuation the code might be misread, as shown above the line. [Modified from Beck, Moffat, and Lloyd (1973)]

cells RNA plays a rather different role, acting as an intermediary between the genetic store and the cytoplasmic sites at which proteins are assembled. This role will be considered next.

Ribonucleic acid (RNA)

Types of RNA

Translation from the triplet code of DNA to the twenty-unit amino acid code of protein molecules is performed by RNA. There are three characteristic forms of RNA.

1. *Messenger RNA (mRNA)*. This is a single-stranded molecule constructed in the nucleus as a complement to one strand of DNA. It has been proposed that the DNA molecule opens sufficiently for an mRNA template to form along an appropriate part of one of the base series, as illustrated in *Figure 81*. The base-pairing rule is similar to that followed during DNA replication, with the exception that the place of thymine (which is not found in RNA) is taken by *uracil*. Thus mRNA *transcribes* the information stored in DNA.

2. *Ribosomal RNA*. The precise structure and function of this, the predominant, type of RNA is still in doubt. It is produced in the *nucleolus*, a specialized region of the nucleus, and is presumably a complement of part of the stored genetic information. Ribosomal RNA is found in the cytoplasm within granules called *ribosomes:* tiny electron-dense bodies found free in the cytoplasm or arranged along the membranes lining the cavities and tubes of the endoplasmic reticulum (*see Figure 82*). Ribosomes can be thought of as the tools with which proteins are fabricated in the cell, and they are the sites where the information carried by mRNA is converted into chains of amino acids.

3. *Transfer RNA (tRNA)*. Molecules of tRNA transfer free amino acids to their correct position on the template provided by the mRNA. For each type of amino acid there is a specific tRNA molecule.

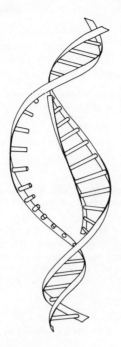

Figure 81. Transcription of genetic information. The two strands of the DNA molecule have separated sufficiently to allow a molecule of mRNA to be built along the exposed bases of one of the strands.

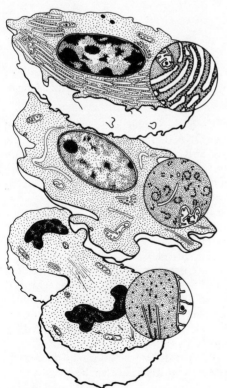

Figure 82. Distribution of ribosomes. Three cellular states are shown: a dividing cell *(bottom),* an undifferentiated embryonic cell *(middle),* and a differentiated cell *(top).* Small portions of the cytoplasm are drawn at higher magnification to show differences in the distribution of ribosomes *(circular insets).* In the dividing cell the ribosomes appear as evenly distributed granules dispersed between other organelles; in the embryonic cell they form characteristic clusters or 'rosettes' along strands of mRNA; and in the differentiated cell most of them are attached to the membrane system of the endoplasmic reticulum – free ribosomes are rare.

Assembly of proteins

Before free amino acids can be built into proteins they must first be activated. This is achieved when the amino acid reacts with adenosine triphosphate (an energy-rich molecule) and a specific enzyme. The result is an activated complex of amino acid/ adenosine monophosphate/ enzyme, which can then combine with an appropriate molecule of tRNA. During combination the activating enzyme and adenosine monophosphate are lost. Finally, the amino acids are assembled into chains with the order specified by the mRNA. *Figure 83* shows how this is done. The ribosomes move along the mRNA molecule from one end to the other, matching the tRNA/ amino acid units to the base triplets. Several ribosomes can 'read' one molecule of mRNA at the same time by moving along in succession, separated by small intervals. This can often be seen in electromicrographs of embryonic cells in which rapid protein synthesis has begun; ribosomes that were originally scattered evenly throughout the cytoplasm appear as clusters and rosettes, with a strand of mRNA linking them. During cell division, however, the ribosomes become evenly distributed once more, suggesting a temporary pause in synthesis. In differentiated cells, free ribosomes are less common and most synthetic activity occurs along the membrane system of the endoplasmic reticulum (*see Figure 82*).

Messenger RNA molecules do not function indefinitely, and it is currently estimated that the template can only be used ten to twenty times before it becomes nonfunctional.

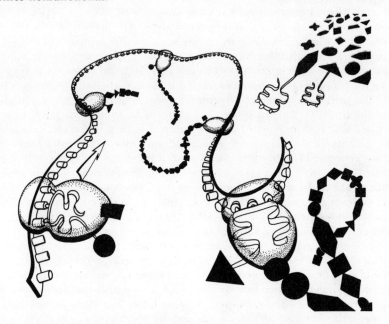

Figure 83. Protein synthesis. In this diagram five ribosomes are 'reading' the same molecule of mRNA. Amino acids from the cytoplasmic pool *(upper right)* are attached to specific tRNA molecules, which then match the amino acids with the triplet codons of the mRNA. In this way a chain of amino acids is assembled in the correct sequence to form a particular protein.

With this understanding of the mechanism of protein synthesis, it became possible to explain the action of certain antibiotics that are known to produce malformations in embryos of experimental animals. It was shown that actinomycin blocks the synthesis of new mRNA in the nucleus, while chloramphenicol competes with mRNA for sites on ribosomes. Thus the observed malformations stem from either a breakdown or a retardation in protein synthesis.

Gene expression

So far we have described how information is stored within the nucleus and the mechanism by which this information can direct the manufacture of specific proteins in the cytoplasm. We have still to explain why some cells differentiate in one way while other cells in the same body, and thus containing the same stored information, manufacture a different complement of proteins and become cells of other types. The problem is one of selecting different sets of information from a common pool, and so far the most cogent explanation of how this may occur is the operon concept.

The operon concept

Jacob and Monod postulated that chromosomes carry three types of genes: *structural* genes, *operator* genes, and *regulator* genes. In this model (illustrated in *Figure 84*), structural genes direct the synthesis of specific proteins via mRNA, but their activities are controlled by an operator gene. One or more structural genes together with the associated operator gene constitute an *operon*. It is suggested that the functioning of an operon is related to the metabolic requirements of the cell by the third type of gene: a regulator gene. The regulator gene is able to repress the operator gene in some way, perhaps by synthesizing RNA *repressor* molecules. The operator gene, when combined with a repressor, is unable to induce activity in the structural gene, and the operon is then *repressed*. The regulator gene itself can become inactivated. For example, it is possible that the substrate of a particular enzymic reaction can combine with a repressor molecule and inactivate it. In this way the operon would become *derepressed*. Thus the operon concept suggests a feedback mechanism from the cytoplasm to the nucleus and provides the means by which the stream of information flowing from the nucleus can be modified to suit the immediate needs of the cell.

This raises the question as to whether embryonic induction occurs through a modification of this mechanism. It has been proposed that inducers activate or inactivate specific regulator genes, thus modifying protein synthesis in the cell. Alternatively, it is suggested that specific mRNA molecules pass from the inducer to the reactive tissue. Some support for this latter view was indeed obtained experimentally. Autoradiographic studies of the developing eye in chick embryos showed that the concentration of RNA was originally higher in the optic vesicle than in the overlying ectoderm, but that during the period of induction this relationship became reversed. However, it was not possible to show that an actual transfer of molecules had occurred rather than a change in the patterns of RNA synthesis in the apposed tissues.

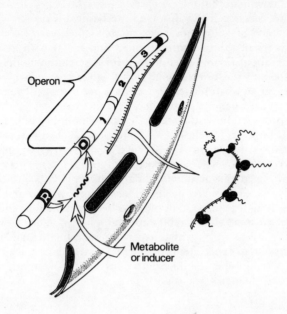

Figure 84. Regulation of protein synthesis: the operon concept of Jacob and Monod. Part of a chromosome is shown *(left)*, which carries three types of gene: a regulator gene (R), an operator gene (O), and three structural genes (1, 2, 3). The operator and structural genes together constitute an operon. The operon is held dormant by a repressor molecule (wavy line) produced by the regulator gene. When the information contained in the structural genes is required, the repressor molecule is inactivated and the operator gene allows an mRNA template (fringed line) to be made of the information stored in the structural genes. The mRNA then leaves the nucleus through one of the pores in the nuclear membrane *(centre)* and enters the cytoplasm *(right)* where protein synthesis can begin.

Model Building

The use of models in scientific research

The avenues of research discussed above can all be studied directly and experimentally. However, scientific research consists of more than the random gathering and interpretation of experimental results. There must first be the selection of suitable questions for investigation — questions that arise from our present incomplete understanding. In formulating questions or hypotheses it is necessary first to make an imaginative or predictive step forward and then to test the validity of the suggestion by experiment. Is the hypothesis a good approximation or is it false? (Note that the word 'true' or 'proven' cannot safely be used in science, particularly biology, since observations are generally based on only a small sample of the population being studied. At best, we can say that 'so far, we have no reason to doubt that ... '. Similarly, it is quite impossible totally to disprove an hypothesis experimentally.) Thus scientific progress relies on creative thought, a comment that may surprise those who believe that only the Arts are creative while science grows in a cold mechanical way.

An important facet of contemporary research is called *model building*. This term conjures up images of things made of wood and metal with a hint of Heath Robinson charm about them, which indeed some scientific models have, but in the present context the meaning is much wider. Models can take many forms, and many exist only in an abstract sense, e.g. as a mathematical relationship or a computer program. Two models have already been mentioned in this section: the model of DNA proposed by Watson and Crick, exemplifying a 'hardware' model that can actually be built, and the operon concept of Jacob and Monod, which has a more abstract form. In general, the aim of a model is to assemble in a coherent way what is currently known about a specific subject. It is often necessary to make assumptions, where current understanding is incomplete, before the model can be used. When a suitable model has been formulated, predictions can be made about the real system (i.e. the system that has been emulated by the model) from the behaviour of the model as variables are manipulated. Thus a model can help to clarify thinking and suggest which aspects to investigate further. The results gained from subsequent experiments on the real system can be used to reappraise the model, either reinforcing it if the original predictions were correct, or helping to modify it if discrepancies are apparent.

Cybernetics

In the late 1940s Norbert Weiner introduced a new word — *'cybernetics'* — to describe one aspect of this scientific modelling approach, defining it as 'the science of control and communication in the animal and the machine'. His intention was to bring together information and concepts from many different disciplines in order to clarify understanding in biology and technology, particularly in relation to how collections of things can work together as viable functioning systems. He emphasized that similarities exist between biological systems and man-made systems and suggested that it would be valuable to extract the basic principles at work in viable systems of either type. The similarities between animate and inanimate or man-made systems encourage the study of models that at first sight seem to bear little direct comparison with the real system, but that nonetheless are organized, hopefully, on the same basic pattern.

The application of cybernetic modelling methods to the study of development may bring rich rewards in view of the significance of organization and control during development. Many models have already been proposed, some mechanical, some in more abstract form such as computer programs, and they have been variously directed towards individual events in the flow of development or more general problems such as the phenomenon of self-reproduction. All the models have in common the basic hypothesis that communication between the subunits of the system must occur, and for this reason studies of *automata nets* have become popular.

An automata net consists of a set of units, each of which responds in a predictable fashion to a particular input. The units are interconnected so that each automaton can communicate with at least one other in the net, and thus there is a resemblance to the grouping of cells to form an organism. These basically simple models have already demonstrated that patterns – i.e. spatial differentiation – can arise even if no apparent pattern existed in the network before it was activated.

Information theory

The phenomenon of communication has itself received close scientific study, and the observations have been brought together in *information theory*. This embodies a number of principles that provide a precise and mathematical way of describing and analysing communications systems. To generalize, information can be thought of as meaningful messages conveyed through a communication channel from a transmitter to a receiver. A unit of information that is often referred to is the *bit*, which is defined as the quantity of information required to determine between two alternatives with equal probabilities, e.g. yes/no. The message, in most cases made up of a number of bits of information, has generally to be encoded before transmission and decoded at the receiving end. Ideally, none of the content of the message is lost during coding, transmission, and decoding, but in practice the efficiency of the system is reduced by interference or *noise*. To compensate for the loss of meaning produced by noise, all or part of the message may have to be repeated, thus introducing a degree of redundancy.

Information theory is now being used in biological studies. For example, it is clear that the mechanism of protein synthesis lends itself to analysis by information theory, with the chromosomes playing the part of a transmitter, the mRNA providing a communication channel, and the ribosomes helping to convert the coded message into a protein molecule. The statistical techniques that have been developed may also help to resolve the age-old controversy of preformation versus epigenesis. Attempts have been made to assess the amount of information in organisms as they develop to see if there are any measurable increases or decreases. Some estimates are based on the number of atoms or molecules in the organism, in which case it is possible to argue that the adult contains more information: 10^{25} bits compared with 10^{15} bits for the mammalian egg. However, conclusions based on numbers of atoms or molecules are probably of limited relevance in a consideration of development, since organisms are open systems absorbing raw materials and energy and excreting waste products, and thus molecules are being continuously added and subtracted. It would be more interesting to quantify changes in complexity and organization, concentrating only on situations where new information is expressed rather than including in addition those processes where an existing set of instructions is repeated over and over again, as happens in structures composed of a large number of identical units.

Storage and transfer of developmental information in the cell

Although a reasonable estimate can be made of the number of bits of information carried by DNA in a cell, it is obviously far more difficult to quantify the amount of information needed to control development of an individual. Early attempts have provided a number of different answers, but interestingly the figures quoted are all substantially larger than the figure for nuclear DNA. This strongly suggests that epigenesis occurs, but there could be another explanation. For example, it is possible that the zygote contains information in addition to that stored in the nucleus. There is indeed evidence that the cytoplasm of the fertilized egg contains information that is essential to at least the early stages of development. It seems that the nuclear genes play little part in the early cleavage divisions and only later exert an influence on cellular activity. This was shown by the following series of experiments.

When the nucleus of a fertilized egg of an amphibian was removed by sucking with a micropipette and the nucleus from a differentiated cell of the same species was put in its place, development generally proceeded through the early stages and then stopped. Only nuclei from intestinal cells of adult amphibia could support later development. However, if the original nucleus was replaced by the nucleus from a fertilized egg of another species of amphibian, development proceeded through the early stages as before and then continued, although the patterns of differentiation were largely dictated by the genetic information of the implanted nucleus. When the nucleus of a fertilized egg was placed in an enucleated, but differentiated, cell of the same species, development did not occur. The results of the experiments were not quite as clear-cut as this summary suggests, but they illustrate a number of interesting points:

1. The ability of early development to occur even if an inappropriate nucleus is present.
2. The irreversible loss of potential that occurs in the nucleus of some cells as they differentiate.
3. The importance of nuclear information for the development of species-specific characteristics.
4. The inability of a nucleus from a fertilized egg to initiate development when incorporated in differentiated cytoplasm.

In more general terms, the experiments beautifully illustrated the significance of the interaction between the nucleus and cytoplasm during development.

In the early cleavage divisions there is subdivision of the original egg cytoplasm into an increasing number of cells, and at first there is no synthesis of new cytoplasmic materials. Thus any differences in the initial distribution of various cytoplasmic components would result in a gradual development of differences between adjacent cells as cleavage proceeded, as shown in *Figure 85*. It is tempting to speculate that this increasing asymmetry would initiate different patterns of gene expression in different regions of the conceptus and trigger the various sequences of differentiation.

There are several sites where cytoplasmic information may be stored. The abundant free ribosomes may carry developmental information, and it has been suggested that small amounts of cytoplasmic DNA exist as *cytogenes*. Mitochondria also contain DNA (*see Figure 86*), but it is probably used only by the mitochondria themselves as they grow and reproduce. Incidentally, this partial autonomy of the mitochondrion, together with its distinctive form, has led to the proposal that primitive cells at a very early stage in evolution were infected by a bacterium-like organism. It is thought that this organism established such a successful symbiotic relationship with the cell — receiving nutrients from it and giving back energy-rich molecules — that the relationship has persisted.

Theoretical Aspects

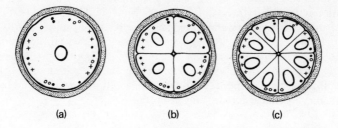

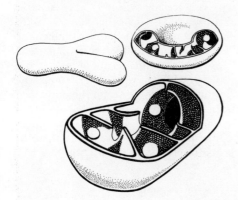

Figure 85. Subdivision of the cytoplasm of the zygote during cleavage divisions. Cytoplasmic components are represented by circles and crosses. The cytoplasm of the zygote (a) is subdivided to form comparable cells at first (b), but a stage will be reached where the cytoplasm differs from cell to cell (c) and symmetry will be lost.

Figure 86. Mitochondrial form and structure. Parts of the envelope have been removed from two of the mitochondria to reveal their internal structure. The wall consists of two layers, and the inner layer is infolded to form partitions, or cristae. Arrays of enzyme molecules cover the internal aspect. The large white spheres floating in the central cavity represent aggregates of mitochondrial DNA.

It does not seem that the spatial arrangement of cytoplasmic components in the egg is fixed for a given species, or that a precise pattern of distribution is essential for normal development. This was shown experimentally: it was found that redistribution of cytoplasmic contents by centrifugation does not prevent normal development.

Inadequacies of the models

Even if allowances are made for this phenomenon of cytoplasmic storage of developmental information in the egg, there still remains a significant discrepancy between the information available at the start of development and the amount apparently needed during development. This may of course be an artefactual difference stemming from inappropriate methods of calculation, but if the difference is real we can only conclude that we are approaching the study of development wrongly. It is obviously satisfying to imagine the stored DNA precisely controlling each step during development, like a craftsman guiding his tools, but could it be that we are imposing our own technologically biased way of thinking on a very different type of system? Cyberneticians do not suggest that all systems are alike, only that certain types of organization can be seen in both

146 *Before Birth*

biological and inanimate systems. Therefore discretion must be used in interpreting the few clues we have that relate to biological development. For 1 500 years the study of cosmology stagnated because of stubborn belief in a machine-like system of spheres within spheres, perfect circles, an earth-centred universe. Perhaps we have a similar deadlock in embryological thought now, becoming obsessed by the details of isolated events rather than first looking for patterns in the overall flow of development. In the next subsection we will stand back a little and take a more general look at development.

Developing a Model of Development

We have looked briefly at some of the existing models of development. Why not formulate one of our own? The materials required are simple: a reasonable working knowledge of the subject as it currently stands, a little imagination, and a supply of paper on which to note down ideas. No one can yet claim to understand fully how development occurs, so this is a very rewarding and fertile field in which to work. A new model may not ultimately provide anything useful or lasting, but at least it may help to organize the rapidly growing mass of observations and encourage more critical thought, and at best it may become of significant value.

As we have seen, some models concentrate on one small problem and explore it in detail, but we shall adopt a different approach and try to build a more generalized model of the flow of development. This aim may turn out to be overambitious, but that should not deter us from making the attempt.

Preliminary steps

Questions to be answered
First, we must formulate a few questions that we should like the model to tackle and perhaps answer. In this way we begin to select a particular model from the presumably infinite number possible; we establish limits within which to work. For example, we can ask:

1. Why does development usually flow in one direction?
2. What is the motivation behind development?
3. Does the zygote contain, or need to contain, information to direct *all* the changes that occur during development?
4. Why do cells containing identical stores of genetic information eventually differentiate along different paths?
5. Is cancer really a reversal of differentiation?

Note that the questions given here are controversial rather than rigorously critical at this stage.

Relevant observations
Second, it is helpful to note down in shorthand the research observations that are relevant to our general model. These cannot initially be placed in any special order since we do not know the relative significance of each to our model, but the

Theoretical Aspects 147

list will make a useful reference point as we continue to design and modify. The list is unlikely to be complete from the first, but we can supplement it later if necessary.

1. Early embryonic cells show great potential, but this potential diminishes as differentiation progresses.
2. Development is triggered by fertilization, but parthenogenesis may occur in ova.
3. Some embryonic regions regulate the development of other regions.
4. Each organ or system passes through critical period(s) when environmental factors may have a profound teratogenic effect.
5. Cancer cells are structurally simple, capable of multiplication, and apparently immune to external control mechanisms.
6. A single cell (zygote) is converted into a complex adult with many cells.
7. Cleavage does not appear to be under nuclear control.
8. The nucleus (not only the cytoplasm) may become modified during differentiation.
9. The ability to regulate or regenerate becomes less as complexity increases.

Starting the ball rolling

Once this patchwork of ideas is established, the next step is to allow the ideas to mix and interact until some unifying pattern emerges. This is the most unpredictable part of model building. It may take seconds, days, or years to obtain some kind of starting point, or perhaps nothing will take shape, in which case it is necessary to find more information with which to work. Often ideas from apparently unrelated sources can act as the necessary trigger. For example, the model outlined in the following pages took shape during a discussion with a physicist, and it tries to profit from significant changes in thought that have occurred in that area, but suitable ideas might equally have arisen from consideration of phenomena as diverse as development in human societies or changes that occur during the evolution of a language.

A model based on developmental potential

Assumption: developmental potential is comparable to potential energy

Let us propose that the key factor in development is the relationship between developmental potential and differentiation. The zygote, apparently at least, is structurally simple and yet contains enormous potential. On the other hand, a cell taken from an adult has a potential that is inversely related to its degree of specialization — the more it is differentiated, the less developmental potential it has — and we can generalize that the mean potential of all the cells in an adult is probably far below that of the zygote. This is graphically illustrated in *Figure 87*. Thus it seems that structural and functional complexity is gained at the expense of potential. Let us assume, then, that potential acts in some way as a motivation for development, a stored 'developmental energy'. In the same way that a stone pushed from the top of a smooth hill will roll to the bottom, expending some of its potential energy to overcome friction and converting another component into momentum, so will a zygote with sufficient potential (it is hypothesized) begin to descend a slope of differentiation, losing potential but gaining in complexity.

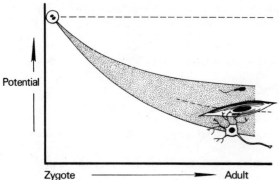

Figure 87. A comparison between the potential of a zygote and the mean potential of the tissues of an adult. It is hypothesized that some tissues lose more potential than others as they differentiate, and thus different cell types retain different levels of potential. However, it is proposed that the mean potential for all the adult tissues is considerably less than that for the zygote.

The hill analogy

Waddington suggested this hill analogy some twenty years ago. He proposed that the side of the hill is channelled by a number of valleys radiating away from its crest, each valley becoming deeper towards the foot of the hill. Thus a cell starting from the summit can choose (or is pushed towards) one of the valleys before it, but once it has entered a valley it has become committed to a particular 'fate', or type of differentiation, and cannot turn back. *Figure 88* illustrates the analogy.

Let us extend this analogy to include not just the differentiation of an individual cell but also the overall sequence of development. It has been noted that under normal circumstances a single gamete cannot develop into a new individual. Hence it could be that fertilization is a means of raising potential above a 'threshold' by summation of the potential possessed by the two gametes so that the developmental descent can begin (*see Figure 89*). In terms of the analogy, fertilization pushes the cell to the top of the hill. However, the phenomenon of parthenogenesis, in which an ovum begins to develop without the penetration of a spermatozoon, suggests that the ovum already possesses a high potential and that even mechanical stimulation may lift it above the hypothetical threshold, if only sufficiently for a brief period of development (*see Figure 90*). In contrast, spermatozoa have shown no ability to undergo parthenogenesis and may therefore be considered to contain less potential than the ovum.

Notice that the term 'potential' is used in an entirely abstract sense in this discussion. There is no need at present to try and explain it as being a consequence of nuclear DNA or to try and locate where the additional potential of the ovum resides. In this model we are more concerned with basic patterns than with mechanisms. (If we wanted to understand how an aircraft flies, it would be rather foolish to begin by dismantling the flight deck, however important it is to the aircraft as a whole. It would be more profitable to observe first the overall behaviour of the aircraft.)

It seems likely that the gametes contain more potential than any other cells in the adult body, and in view of the early segregation of their precursors, the primordial germ cells, we can propose a much shallower developmental slope for them than for the remainder of the individual (*see Figure 91*).

Already we can envisage differing rates in the loss of potential as cells differentiate during development, and the contours of our developmental hill must reflect this, becoming more complex in later stages, i.e. lower down the slope. But,

Theoretical Aspects   149

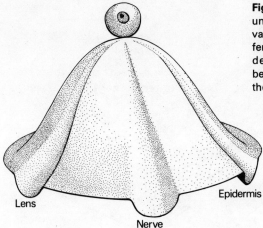

Figure 88. The developmental hill. An undifferentiated cell faces a number of valleys, each of which leads to a different fate for the cell. As the cell descends it loses potential and becomes increasingly committed to the valley it has entered.

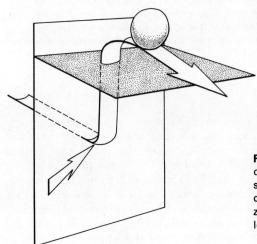

Figure 89. The effect of fertilization on developmental potential. It is suggested here that the potential of the cell resulting from fertilization – the zygote – is raised above a threshold level so that development can begin.

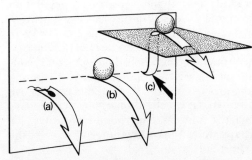

Figure 90. Changes in the development potential of gametes. Usually a spermatozoon (a) or an ovum (b) degenerates soon after release if fertilization does not occur. However, an ovum may, if appropriately stimulated, begin to develop parthenogenetically (c) – although the gain in potential may only be sufficient to allow a brief period of development before the embryo sinks below the hypothetical threshold.

150 Before Birth

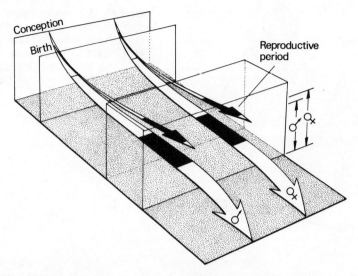

Figure 91. Developmental curves. It is suggested in this diagram that the developmental slopes for the gametes (slender arrows) are shallower than the slopes for the remainder of the individual (broad arrows). As shown here, it is possible that the female gametes retain more potential than their male counterparts.

of course, we must remember our original premise that it is the hill that guides the cells, not vice versa.

How can we display the changing patterns of potential graphically? *Figure 92* attempts to do this. It can be seen that the valleys form a diverging system and that there are differences in slope both along a particular valley and between different valleys. This is to underline common observations that the rate of development varies from stage to stage and from region to region. The phenomenon of induction has also been represented, being shown as a confluence of two (or more) valleys where cells collide as they descend and are deflected from their original paths into different valleys. Cell death, another important facet of normal development, is suggested by pathways that drop almost vertically to zero potential.

A limitation of the model

This idea of a conceptus rolling down a hillside, with the ever-growing population of cells channelled into a system of diverging valleys, is quite pleasing. It helps to link the concept of falling developmental potential with increasing complexity of structure and function. However, one possible weakness can be seen in *Figure 92:* the hill stands rigid and unchanging throughout the whole developmental sequence, rather like a mountain that supports a season's skiing without becoming appreciably altered. Is it reasonable to propose a static slope of this type, or are there better alternatives? Before continuing, it is worth summarizing a cautionary tale provided by another branch of science: physics. Sadly, embryology lags far behind physics in maturity, but perhaps we can take advantage of some of the physical concepts and use them as short-cuts to our goal.

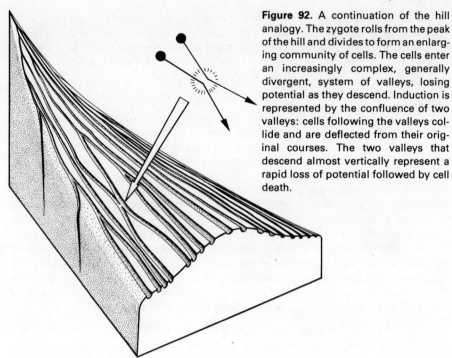

Figure 92. A continuation of the hill analogy. The zygote rolls from the peak of the hill and divides to form an enlarging community of cells. The cells enter an increasingly complex, generally divergent, system of valleys, losing potential as they descend. Induction is represented by the confluence of two valleys: cells following the valleys collide and are deflected from their original courses. The two valleys that descend almost vertically represent a rapid loss of potential followed by cell death.

A lesson from physics

Theoretical and experimental physicists have come to accept that some of the apparently fundamental properties of our universe — space, time, and gravity — do not conform with our commonsense expectations of them. Naturally, we are acclimatized to living in a particular region of space at a particular time, and our sense of scale is largely determined by our own size and life span. Many of our scientific attitudes are related, often unconsciously, to these rather narrow limits.

Space and time

For a long time scientists accepted the Newtonian ideas of space and time, formulated in 1687: 'Absolute space, in its own nature, without relation to anything external, remains always similar and immovable. Absolute, true, and mathematical time, of itself, and from its own nature, flows equably without relation to anything external.'

However, this belief in ideal rigid frameworks of space and time no longer exists. Contemporary physicists often find it difficult to separate space from time and prefer to think of a combined phenomenon they call *spacetime*. To quote Minkowski (1908): 'Henceforth space by itself, and time by itself, are doomed to fade away into mere shadows, and only a kind of union of the two will preserve an independent reality.'

The need for this composite view of space and time arose largely from the work of Einstein, who showed that physics is simple — i.e. obeys the Newtonian laws of motion and can be expressed by Euclidean geometry — only when considered in a

.small region of space and time, and that we are unable to apply the techniques of Euclidean geometry to the analysis of spacetime because there seems to be an additional dimension or *curvature*. Physicists have to resort to a more flexible type of geometry — Lorentzian geometry — when discussing, for example, astronomical events.

Gravitation

The next important point to be aware of is that the curvature of spacetime affects the motion of particles. This effect of geometry on matter corresponds to the classical idea of gravitation. If we study the path of a particle over a small period of space and time we observe a straight line, but over more extended regions of space and time the path appears curved and may begin to approach other paths. The rate of convergence is governed by the degree of curvature of spacetime. The distinction being made here between classical and current ideas of gravitation is that matter is now thought to respond to *local* conditions, or curvature, of spacetime, rather than to be influenced by aggregations of matter at a distance. Clearly, the distinction is one of convenience — it provides a useful way of simplifying the study of events — and does not imply that local curvatures of spacetime are unrelated to the presence of matter elsewhere. Indeed, the presence of matter produces changes in spacetime — it warps the geometry — and this two-way relationship may be summarized in the following way.

> Spacetime acts on matter, telling it how to move.
> In turn, matter acts on spacetime, telling it how to curve.

Waterbed analogy

Perhaps we can begin to visualize the modern concept of a gravitational field with the help of an analogy: a water-filled bed. The upper surface of a waterbed is initially flat or perhaps gently curved. A ball bearing will roll smoothly across the surface in a straight or gently curving path, producing negligible distortion of the plastic membrane. However, if a heavy object is placed on the surface the membrane is distorted to form a hollow. The heavier the object, the larger the deformation produced. If a ball bearing is now rolled across the bed its path will be influenced by the deformation in the surface. As it rolls down into the hollow it accelerates and, depending on its original path, may collide with the heavy object. If more and more stationary weights are placed on different parts of the surface of the waterbed, it can be imagined that the path of the ball bearing will become increasingly complex as it is deflected by the crests and hollows of the surface, as *Figure 93* illustrates. Its velocity and direction will vary continuously. Now let us imagine that the heavy weights begin to move in relation to each other — vertically, horizontally, or both — so that the patterns of deformation are constantly changing. An additional complication to these patterns will be caused by ripples radiating out across the surface from oscillating weights. Clearly, under these conditions the path of a rolling ball bearing will be very complex indeed, and an analysis of its motion will be simple only over small portions of its journey (physics is simple only when the region of spacetime studied is small enough to look flat).

We can see in this waterbed analogy that matter influences the curvature of the membrane, while in turn the curvature of the membrane governs the motion of matter. But it should be remembered that many simplifications have been made.

Theoretical Aspects 153

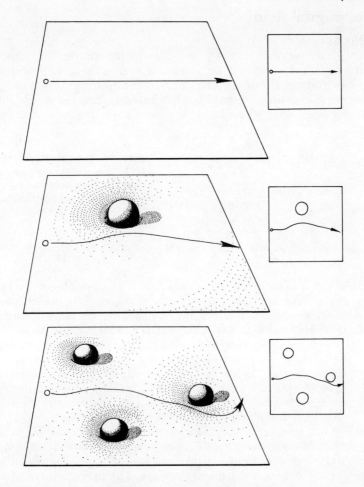

Figure 93. The waterbed analogy. The path taken by a ball bearing becomes more complex if the surface of the waterbed is distorted by the presence of heavy objects.

For example, we have considered only a single membrane linking the various events. Next we should try to imagine a three-dimensional array of membranes arising from each lump of matter and extending to merge with membranes from other bodies. In some way an event occurring on any particular membrane will produce changes throughout the system, although we may anticipate that the effect diminishes with increasing distance from the source and that there will be interaction with other changes.

Thus we can perhaps come a little closer to understanding the relationship between space and matter, provided we can finally dispense with the idea of membranes, which of course do not exist in gravitational systems, and with other aids to our imagination that have no place in reality. But how can all this help us with our model of development?

Developmental field

The field concept

Let us replace the mental image of a fixed hill with the idea of a flexible sensitive membrane or *field*. In addition, let us assume that as the field provides information to the embryo, the developing embryo has a reciprocal effect on the field and changes it. Let us imagine that when fertilization occurs the membrane takes the form of a single peak whose height is related to the potential of the cell (*see Figure 94*).

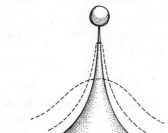

Figure 94. The developmental field after fertilization. A single narrow peak supports a high-potential zygote. It is proposed that the field is unstable and begins to flatten and expand (dashed profiles).

We can perhaps assume that this initial field is an unstable one and rapidly begins to change. An approximate idea of a changing field can be gained by placing a tablecloth or sheet on a flat surface and pulling the centre up into a peak. When the material is released the peak flattens quickly, and as the trapped air escapes the sheet sinks back slowly to its original position, although rather more wrinkled than before. A careful observer may notice some wave-like ripples spreading across the sheet as it sinks (reference will be made to ripples later).

Returning to the zygote, we can envisage a rapid fall in potential (flattening of the peak in the field) as cleavage occurs, although at this stage there is little change in structural complexity. Then, as the cell population grows and more potential is lost, a number of irregularities develop in the expanding field. As a result some cells begin to lose potential more rapidly than others (*see Figure 95*).

An important point to notice is that our proposed developmental field is changing continuously as the embryo develops. The field, as it contributes to development, becomes changed by it, so that there can be no turning back. This is the major difference between the hill analogy and the field concept.

Thus, with the passage of time the landscape dominated by a single dramatic peak becomes steadily flatter, but concurrently the irregularities of its surface become more complex and subtle.

Disturbances to the field

It has been suggested that the field is sensitive. This means that when a disturbance is generated somewhere in the field its effect will be communicated to the remainder of the field. This type of communication may play a part in co-ordinating the development of each region in relation to the whole individual. Could this also be a helpful way to envisage the action of some teratogens? Perhaps a teratogen can indent, or tilt, or in some other way modify the developmental field so that development begins to go astray. A mild insult may only produce a transitory ripple of disturbance from which the embryo can recover, whereas a more severe disturbance may completely upset the balance of the field and cause development to become irreversibly abnormal.

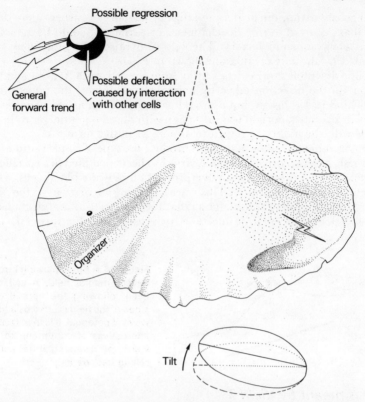

Figure 95. The developmental field at a later stage. The field *(centre)* is more expansive and shows many irregularities. The central peak is much flatter than the peak at fertilization (dashed line). A large radial valley is included to represent an organizer region of the embryo. The impact of a specific teratogen is represented by a deeply distorted region in the field (centre right), while the effect of a general teratogen is shown separately as a tilting of the entire field *(lower right)*. *Upper left:* An enlarged view that summarizes the factors influencing the movement of a particular cell.

Ripples in the field

How can we relate the phenomenon of cancer to the field hypothesis? It has been noted earlier that cancer cells tend to have a simple physical structure, and this, together with the attribute of rapid uncontrolled cell division, suggests that they may have regained at least some of the potential lost during development. However, tumours that develop later in life rarely contain a mixture of differentiated cells: the cells usually remain primitive. This may indicate that potential alone is not enough for development: a suitable field must also exist.

Perhaps a rise in potential could be produced by a ripple passing through the developmental field. Consider a stage at which the definitive structure and function of the individual have been established. Conceivably a ripple in this otherwise quite stable field could raise the potential of a cell (or group of cells) just sufficiently to initiate the change we call cancer. Such a ripple might arise spontaneously within the field as a vestige or memory of earlier, more dramatic events that occurred during development, or perhaps it could be generated by external environmental factors. The cell with raised potential may now be confronted by the sort of situation shown in *Figure 96*.

A gentle downhill slope extends away from the cell, both in front and behind, one slope in the direction taken by normal development, the other extending away behind it. In this situation the cell is delicately balanced between rolling forward in an almost normal way, although with raised potential, or rolling back, perhaps with the ability to return to a more primitive form.

An environmental factor that might produce the type of ripple postulated here is, of course, a virus. Already an RNA virus has been identified in a specific type of leukaemia (which is basically an overproduction of white blood cells) and may represent an underlying cause. This research, being carried out at the National Cancer Institute, Maryland, suggests that the virus acts as an infectious agent, although it does not seem to be infectious in the usual sense because of the rarity of the disease.

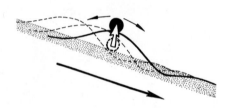

Figure 96. A ripple passing through the developmental field. A cell that has been following the normal downhill slope of the field is lifted by a transitory wave of potential. It is then faced by the alternatives of continuing to roll forwards or reversing its direction and rolling backwards.

What is meant by 'field'?

The word 'field' has been used up to this point without any attempt to define what form it takes, although a general allusion to a gravitational field has been made. Similar field concepts have arisen before in the context of development. In 1912 Spemann suggested that each organ develops in an area analogous to a diffraction circle, the degree of 'determination' or disposition to form the structure being highest in the centre and decreasing peripherally. Harrison, another eminent embryologist, echoed this view several years later. However, in 1941 Child warned that:

> 'Without further analysis the field concept, as applied to development, has only a formal value . . . reference to a field merely states experimental data in terms of an unknown, of a concept without definite content, and the field often becomes little more than a verbalistic refuge.'

He went on to discuss the importance of physiological gradients during development:

> 'The experimental evidence concerning physiological gradients seems to indicate that developmental fields in their simplest, most general forms are gradient systems, that is, the field is constituted by the gradient or gradients present; the gradients are the vectors of the field and determine its extent and orderly relations within it.'

More recently, Gierer has developed a computer model of pattern formation in the simple coelenterate, hydra. As a basis he used the interaction of two gradients: one with an activating influence, the other with an inhibiting influence. He found that quite complex three-dimensional patterns could arise.

Theoretical Aspects

An alternative explanation is that the field consists of certain inevitable chemical and physical tendencies: the sum of the changes or interactions that can occur without genetic supervision. As an illustration, imagine a stone thrown into the air. It curves gracefully away, falls to the ground, bounces, rolls, and finally comes to rest. The stone follows quite a complex path through space and time, exhibiting continuous changes in velocity and direction. Could a physicist predict the path before the stone is thrown? Provided he is given details about the stone (density, mass, shape), the way it is thrown (velocity, direction), and the environment (wind speed, type of surface the stone will hit, etc.), he should be able to give an accurate answer. Thus the path is predictable.

In a case like this, where we are considering an inanimate lump of matter, it seems unnecessary to postulate that the stone contains a store of information to guide it along its changing path. It is easy to accept that, given certain conditions at the start and certain properties of matter, the path taken by the stone is quite logical, predictable, and in fact *inevitable*. The path is dictated by the field the stone experiences as it passively follows the local curvatures of spacetime.

Field effects versus genetic effects

Could inevitability play a part in development? Clearly, it would be advantageous. Instead of demanding that every change that occurs during development must be guided by specific instructions stored in the chromosomes, we could suggest that a proportion of the changes occur inevitably: given a state A, which is unstable, then state B will follow as a result of predictable chemical or physical activity, without genetic control.

If some developmental changes can occur in response to field effects, what then will be the role of genetic information? In addition to its role of providing a reference library from which information may be obtained on how to build proteins, its function may also be to override the sequence of field-produced changes when appropriate and thus to divert the flow of development in a specific controlled way. In the example given above state A, it is suggested, tends inevitably towards state B; genetic intervention at some intermediate stage may result in deviation to a state B^1 instead (*see Figure 97*).

With a mechanism of this type we could imagine development – especially the early stages – proceeding without total dictatorship by the genes. Genetic influence would be felt only when it became necessary to make a change that did not follow inevitably from the already existing stage. If this is so, we can begin to see why the fertilized egg seems to carry a rather small complement of information compared with the adult.

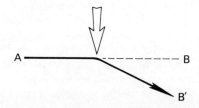

Figure 97. Genetic intervention. State A has a tendency to change towards state B. However, genetic intervention at an intermediate stage may cause deviation to state B'.

Testing the model

We have constructed a field model of development from a blend of physics and biology, and it is now important to devise ways of testing the underlying idea: that a field provides information to the embryo while at the same time becoming modified by the embryo. As noted in the introductory section, we must be careful not to give way to pleasing philosophical ideas that are of no real help in our study.

One method of testing is to apply existing observations to the model to see if any major inconsistencies arise. For example, how does the field idea stand in relation to experiments on induction, on regulation, and on regeneration? Does it help us to understand better our knowledge of abnormal development and of information theory in relation to development?

An alternative method is to design a specific experiment to test a key point of the model, upon which result, ideally, the model will stand or fall. It is obviously difficult and challenging to devise a crucial test, but here is one tentative suggestion, the aim of which would be to determine which changes, if any, can occur without genetic control. Such changes could then provisionally be called *field effects*.

Let us consider first, in the traditional way, a small fragment of development, in this case morphogenesis of the eye. It would be interesting to determine what would happen if component cells were deprived of genetic information at different stages during formation of the lens vesicle and optic cup. Removal of the nuclei would be far too disruptive. A better technique would be to block specifically the formation or transport of mRNA and thus prevent communication between the nucleus and cytoplasm for short periods. It would be important to select a blocking compound whose action is clearly understood (e.g. an antibiotic such as actinomycin), because, as previously mentioned, some teratogens may act indirectly by influencing the field rather than directly by interfering with a specific cellular mechanism. It may be found that abnormal development of the eye occurs regardless of when the dose is given or for how long it acts. This would indicate that nuclear information plays a vital role at all times. On the other hand, it may be found that administration of the blocking agent at some stages produces no measurable effect, while at other stages it can produce abnormal development. A result of this type would suggest that some changes at least can occur spontaneously.

There is already some evidence that invagination of the tip of the optic vesicle is a consequence of invagination of the lens placode — the optic vesicle does not invaginate if the lens placode is not normally positioned in relation to it. Similarly, initial differentiation of the cells of the optic vesicle (towards becoming either neural retinal cells or pigmented cells) also depends on prior formation of the lens vesicle. Current embryological theory would ascribe these changes to the reciprocal induction of cells of the optic vesicle by the cells of the lens placode — reciprocal since at an earlier stage the placode cells were themselves induced by the optic vesicle. Thus the nuclei would be considered to be continuously active and essential. If, however, it was found that these changes can occur, even partially, when the mRNA pathway is blocked, there would be some support for the field idea. A demonstration of this type might also throw light on the apparent discrepancy, according to information theory, between the zygote and the adult.

Final comment

It may seem rather unsatisfactory to end a scientific text with a diffuse mixture of ideas and questions. Undoubtedly some of the ideas given here will turn out to be as ephemeral as the mayfly. However, a mayfly usually lives long enough to ensure the next generation, so perhaps these ideas, although of uncertain permanency, may at least help to generate more ideas. It is important to remember that embryology is not a very mature, and thus highly refined, subject to which it is difficult to contribute. It is still in its infancy and full of doubts and mysteries. Indeed, an entirely false impression would be given if the flaws in current thinking were not emphasized. For this reason the subject has been presented as an inquiry with a somewhat muddled beginning, a rapidly growing present, and a very divergent future. It is hoped that this approach is justified − encouraging rather than disconcerting.

Glossary

The terms used in describing embryonic and adult orientations are illustrated in *Figure 98*.

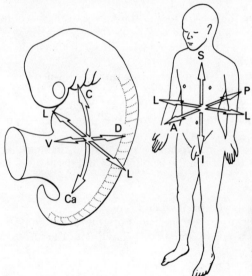

Figure 98. Terms used when describing positional relationships in embryos and adults: C = cranial, Ca = caudal, D = dorsal, V = ventral, L = lateral, A = anterior, P = posterior, S = superior, and I = inferior.

acrosome. A membrane-limited store of enzymes in the head of a spermatozoon.
aetiology. The study of the causes of disease or congenital abnormalities.
alar plate. The dorsal part of the mantle layer of the neural tube. It contains the cell bodies of interneurons and receives a sensory input from the dorsal roots of spinal nerves.
allantois. An endodermal pouch at the caudal end of the embryo. In some species it serves as a receptacle for nitrogenous wastes; later it contributes to bladder formation.
amniocentesis. Extraction of a small sample of amniotic fluid for analysis of the solutes and foetal cells it contains.
amniotic cavity. A fluid-filled cavity, bounded by the amnion, within which the embryo and foetus develop.

atresia. Degeneration of maturing female gametes before they can be ovulated.

autosomes. A chromosome other than a sex chromosome.

basal plate. The ventral part of the mantle layer of the neural tube. It contains the cell bodies of motor neurons whose axonal outputs form the ventral roots of spinal nerves.

blastocyst. The name given to a conceptus just before implantation when it has the form of a hollow sphere of cells.

blastomeres. Cells with high developmental potential produced during the sequence of cleavage divisions that follows fertilization.

blood islands. Collections of mesodermal cells that differentiate to form blood cells.

brain vesicles. Dilated regions of the neural tube at the future head end of the embryo.

bulbus cordis. The most cranial chamber of the primitive heart. Blood passes from it into the arterial system.

cerebrospinal fluid (CSF). A watery fluid produced within the brain. It circulates around the central nervous system before being absorbed into the venous system.

chorion. A general term for the trophoblast and its lining of extraembryonic mesoderm.

choroid fissure. A groove on the ventral aspect of the optic cup through which blood vessels enter and leave the eye.

chromatid. Essentially a chromosome. Two identical chromatids are formed when a chromosome replicates prior to cell division.

cleavage. A series of rapid cell divisions following conception. Cleavage divisions are characterized by subdivision of existing cytoplasm without growth or apparent differentiation.

coelom. A fluid-filled cavity. Intraembryonic coelom: cavity within the embryo; extraembryonic coelom: cavity between the trophoblast and derivatives of the inner cell mass.

conception. See fertilization.

conceptus. A general term for the embryo and its membranes.

congenital abnormality. A structural or functional fault that originates before birth and seriously interferes with the subsequent everyday life of the affected child.

conjoined twins. Incompletely separated monozygotic twins.

corpus luteum. A progesterone-secreting body in the ovary formed from residual cells of a Graafian follicle after ovulation.

critical period. The time at which a developing organ or system is particularly sensitive to harmful environmental stimuli.

cybernetics. The science of control and communication in the animal and the machine.

cyclopia. A single median eye, often associated with other facial abnormalities.

cytogenes. Information stored in clumps of cytoplasmic DNA.

cytotrophoblast. The inner cellular layer of the trophoblast.

decidua. The modified endometrial (uterine) tissue shed at birth.

dermatome. In an embryological context, the part of a somite that contributes to the dermis of the skin.

determination. Irreversible fixation of the developmental fate of an embryonic cell or tissue.

development. Change from simple to complex; the pattern of changes that converts a zygote into an adult.

developmental field. A theoretical concept put forward to explain the emergence of pattern and form during development, utilizing present knowledge about induction and physiological gradients but allowing for other, presently unknown, means of intercellular communication.

differentiation. The process by which cells and tissues become more specialized structurally and functionally. The synthesis of specific proteins is involved, and the process is generally irreversible.

diploid. The normal somatic chromosomal complement. In humans this is forty-four autosomes plus XY sex chromosomes in the male, and forty-four plus XX in the female.

dizygotic twins. Two individuals born to the same mother at the same time and produced by the fertilization of two ova.

DNA. Deoxyribonucleic acid, an information-carrying molecule found in chromosomes.

dorsal root ganglia. Segmentally arranged aggregations of neuronal cell bodies in the sensory dorsal roots of spinal nerves.

ductus arteriosus. A blood vessel that before birth shunts blood from the pulmonary circulation to the aorta. After birth it is transformed into the ligamentum arteriosum.

ductus venosus. A vessel through which blood returning from the placenta bypasses the liver tissue. After birth it becomes the ligamentum venosum.

ectopic pregnancy. Implantation other than within the uterus, e.g. in the uterine tube or peritoneal cavity.

electron microscope. A high-resolution microscope. It operates on similar principles to a light microscope but with an electron gun in place of a light source and magnetic lenses rather than glass lenses. There are two basic modes of operation: (a) transmission electron microscopy, in which thin slices of material are examined; and (b) scanning electron microscopy, in which the shape and surface texture of a specimen are examined.

embryology. The study of the initial stages in an organism's development. In mammals this is mainly centred on the period from conception to birth.

embryonic disc. An oval-shaped plate formed by the embryonic germ layers prior to organogenesis.

endocardial cushions. Elevations that fuse to subdivide the atrioventricular canal in two.

endocardial tube. One of a pair of tubes that fuse to form the primitive heart tube.

endometrium. The lining of the uterus.

epidemiology. The statistical study of the incidence of disease and congenital abnormalities in selected human populations.

epigenesis. The appearance of new structures during development.

extraembryonic. Outside the embryonic body itself.

fertilization. Fusion of a spermatozoon with an ovum to form a zygote.

fontanelle. An area of the skull that remains devoid of bone for some while after birth.

foramen ovale. A valvular opening in the interatrial septum.

gamete. A haploid cell that fuses with another haploid gamete to form a zygote. The male gamete is the spermatozoon; the female gamete is the ovum.

gene. A unit of heredity consisting of a length of DNA capable of specifying an amino acid sequence or regulating the expression of other genes.
germ layers. Three layers of cells: ectoderm, mesoderm, and endoderm, from which all the tissues of the body develop.
Graafian follicle. A multilayered fluid-filled capsule of cells around a maturing oocyte.
gubernaculum. A fibromuscular strand that is believed to assist descent of the gonads.
haploid. Half the normal (diploid) chromosomal complement.
herniation. Protrusion of part of the gut or another abdominal structure into an abnormal site, e.g. physiological herniation of the midgut loop into the umbilical cord.
histones. Proteins that contribute to chromosomal structure.
hydramnios. A greater quantity of amniotic fluid than normal.
hydrocephalus. Accumulation of cerebrospinal fluid within the cranial cavity due to an obstruction at some point in the route by which it circulates.
implantation. Invasion of the uterine lining by the conceptus.
indifferent period. The first 6 weeks of development during which it is impossible to determine the sex of the embryo from its structural appearance.
induction. A controlling influence exerted by one region of the embryo over the development of an adjacent region. The influence is believed to be chemical in nature.
information. This can be regarded as the opposite of uncertainty. Quastler (*Information Theory in Biology,* Urbana, Illinois: University of Illinois Press, 1953) points out that information:

> 'is related to such diverse activities as arranging, constraining, designing, determining, differentiating, messaging, ordering, organizing, planning, restricting, selecting, specializing, specifying, and systematizing; it can be used in connection with all operations which aim at decreasing such qualities as disorder, entropy, generality, ignorance, indistinctness, noise, randomness, uncertainty, variability, and at increasing the amount of degree of certainty'.

inner cell mass. A group of cells on the internal aspect of the trophoblast from which the embryonic body, yolk sac, and amnion develop.
intermediate mesoderm. The column of mesodermal cells lying between the somites medially and the visceral and somatic mesoderm laterally.
intraembryonic. Within the region from which the embryonic body develops.
in utero. Within the uterus.
in vitro. In an artifical environment; e.g. cells in tissue culture are being maintained *in vitro.*
in vivo. In the living animal.
labour. The period of powerful rhythmic contractions of the uterus that culminate in delivery of the baby and its supporting membranes.
mantle layer. A layer of differentiating nerve cells derived from, and lying external to, the neuroepithelium of the original neural tube.
marginal layer. The outermost layer of the early central nervous system composed of processes of nerve cells in the mantle layer. After myelination the marginal layer is referred to as white matter.

meiosis. A sequence of two cell divisions that gives rise to haploid cells, as occurs during formation of the gametes.
membranous labyrinth. A system of fluid-filled chambers and canals that make up the internal ear.
meninges. Three concentric wrappings of the brain and spinal cord: the dura mater, arachnoid mater, and pia mater.
mesonephros. A urine-secreting system found in 6–8-week human embryos.
metanephros. An aggregation of mesodermal cells from which the definitive kidneys develop.
microfilaments. Intracellular fibres, possibly contractile, which are visible by electron microscopy.
microtubules. Intracellular tubules found in mitotic spindles, cilia, flagella, and cells changing shape. They are visible by electron microscopy.
mitosis. The most common type of cell division, in which both daughter cells receive the diploid number of chromosomes.
mitotic spindle. A microtubular array that moves chromosomes apart during cell division.
model. A man-made system in abstract or hardware form that attempts to explain certain natural phenomena.
monozygotic twins. Two individuals that develop from a single zygote.
morphogenesis. The development of form in the embryo.
myelination. The formation of an insulating sheath of concentric lamellae of cell membrane around the processes of nerve cells.
myotome. The part of a somite that differentiates into muscle tissue.
neonatal. Around the time of birth.
neural crest. Migratory cells located initially along the line of fusion of the neural folds. They have diverse fates, becoming nerve cells, Schwann cells, and pigment cells.
neural plate. A thickened region of the ectoderm that is a precursor of the central nervous system.
noise. The loss of information in a transmitted message as a result of outside influences.
notochord. A median rod-like cord of mesodermal cells that probably has a stiffening role in the early embryo and that forms the basis for development of the vertebral column.
oligohydramnios. Less amniotic fluid than normal.
oogenesis. The sequence of changes that converts primitive germ cells into ova.
operon concept. A model put forward to explain the control of protein synthesis.
optic cup. A cup-like structure produced by invagination of the tip of the optic vesicle during development of the eye.
optic vesicle. An outgrowth from the side of the forebrain that develops into the pigmented and neural layers of the retina of the eye.
organization. The association of components into functioning systems.
organizer. A region of the early embryo that plays a significant role in co-ordinating initial stages in development. It is believed to be the site of the first of a series of inductive relationships.
organogenesis. The development of organs.
ossicles. Three tiny bones bridging the cavity of the middle ear.
ossification. The transformation of membranous or cartilagenous tissue into bone.

pharyngeal arches. Bars of mesodermal tissue that give the pharyngeal region of the embryo an appearance reminiscent of the gill region in fishes. The bars are separated by clefts externally and pouches internally.
pharynx. The funnel-shaped tube linking the oral and nasal cavities with the oesophagus and larynx.
placenta. An organ of exchange that supports prenatal development in mammals. It is composed of foetal and maternal tissues and allows approximation of the foetal and maternal bloodstreams.
placode. A thickened plate of cells that usually invaginates to form a pit or vesicular structure.
polar body. A small cell resulting from maturation divisions during oogenesis.
potential. A measure of a cell's ability to give rise to different tissues. Potential is high in early embryonic cells but is lost during differentiation.
preformation. The proposed existence of structures in miniature form in the gametes or embryo that would develop simply by growth in size. This belief lost favour as microscopy improved.
prenatal. Before birth.
primitive streak. A linear thickening of the embryonic disc along which ectodermal cells invaginate to establish the mesodermal layer.
primordial germ cells. Embryonic cells that eventually give rise to the gametes.
pronephros. A urinary system that makes a transitory appearance in 3-week human embryos.
redundancy. Repetition of information to overcome losses due to noise.
regulation. The channelling of the potential of a cell or group of cells along an appropriate developmental pathway during normal development or in response to injury.
replication. The exact copying of DNA prior to cell division.
RNA. Ribonucleic acid, an information-carrying molecule that in mammalian cells is concerned with protein synthesis.
sclerotome. The part of a somite that differentiates to form skeletal tissues.
septum primum. The first partition developed within the primitive atrium of the heart.
septum secundum. The second partition to appear in the embryonic atrium. It co-operates with the septum primum in formation of the foramen ovale.
sex chromosomes. Chromosomes, designated X or Y, that carry sex-determining genes.
sinus venosus. The most caudal chamber of the primitive heart. It receives the venous return from the embryo, yolk sac, and placenta.
somatic mesoderm. Laterally placed mesoderm associated with the ectoderm.
somite. A segmental block of mesodermal cells from which dermal, muscular, and skeletal elements are derived.
spermatogenesis. The sequence of changes that converts primitive germ cells into spermatozoa.
spiral septum. A spiral partition that subdivides the outflow from the heart into the aortic and pulmonary trunks.
surfactant. The collective name for a group of substances that reduce the surface tension of fluid in the lungs and facilitate gaseous exchange.
syncytiotrophoblast. The external syncytial layer of the trophoblast.
syncytium. An aggregation of nuclei and cytoplasm without subdivision into cells.

teratogen. A physical or chemical environmental factor that adversely affects a developing organism.
teratology. The study of abnormal development.
tetralogy of Fallot. A syndrome of four associated abnormalities of the heart.
trophoblast. A sphere of tissues surrounding the developing embryonic body. It is composed of syncytiotrophoblast and cytotrophoblast.
umbilical cord. A twisted cord-like structure linking the foetus with the placenta. It contains blood vessels and remnants of the allantois and vitelline duct.
urorectal septum. A wedge-shaped bar of mesoderm that separates the allantois from the hindgut.
visceral mesoderm. Laterally placed mesoderm associated with the endoderm.
yolk sac. An endodermal sac on the ventral aspect of the embryonic disc. In mammalian embryos it does not contain yolk.
zona pellucida. A coating around maturing oocytes and ova that is composed of glycoproteins and polysaccharides.
zygote. A cell that is formed by fusion of a spermatozoon with an ovum and from which a new individual develops.

Bibliography

Books for further reading and reference

Books marked (A) are advanced accounts. Those marked (G) are suitable for general reading.

- (A) Apter, M. J. (1966) *Cybernetics and Development* (Oxford: Pergamon Press)
- (G) Austin, C. R., and Short, R. V. (eds) (1972) *Germ Cells and Fertilization* (Cambridge: Cambridge University Press)
- (G) Balinsky, B. I. (1970) *An Introduction to Embryology*, 3rd edn (Toronto: Saunders)
- (G) Beck, F., Moffat, D. B., and Lloyd, J. B. (1973) *Human Embryology and Genetics* (Oxford: Blackwell)
- (A) Blechschmidt, E. (1961) *The Stages of Human Development before Birth* (Paris: S. Karger)
- (G) Brachet, J. (1950) *Chemical Embryology* (New York: Interscience)
- (G) Child, C. M. (1941) *Patterns and Problems of Development* (Chicago: University of Chicago Press)
- (G) Davies, J. (1963) *Human Developmental Anatomy* (New York: Ronald Press)
- (A) DeHaan, R. L., and Ursprung, H. (eds) (1965) *Organogenesis* (New York: Holt, Rinehart and Winston)
- (G) DuPraw, E. J. (1970) *DNA and Chromosomes* (New York: Holt, Rinehart and Winston)
- (G) Hamilton, W. J., Boyd, J. D., and Mossman, H. W. (1972) *Human Embryology*, 4th edn (Heffer)
- (A) Kalter, H. (1968) *Teratology of the Central Nervous System* (Chicago: University of Chicago Press)
- (G) Langman, J. (1975) *Medical Embryology*, 3rd edn (Baltimore, Maryland: Williams and Wilkins)
- (A) Morison, J. E. (1970) *Foetal and Neonatal Pathology*, 3rd edn (London: Butterworth)
- (G) New, D. A. T. (1966) *The Culture of Vertebrate Embryos* (Plainfield, New Jersey: Logos Press)
- (A) Parkes, A. S. (ed.) (1966) *Marshall's Physiology of Reproduction*, 3rd edn (Harlow: Longman)
- (G) Patten, B. M. (1968) *Human Embryology*, 3rd edn (New York: McGraw-Hill)
- (A) Romanoff, A. L. (1960) *The Avian Embryo* (London: Macmillan)

(A) Spemann, H. (1938) *Embryonic Development and Induction* (New Haven, Connecticut: Yale University Press)
(G) Trinkaus, J. P. (1969) *Cells into Organs* (Englewood Cliffs, New Jersey: Prentice Hall)
(G) Waddington, C. H. (1940) *Organizers and Genes* (Cambridge: Cambridge University Press)
(G) Waddington, C. H. (1956) *Principles of Embryology* (London: George Allen and Unwin)
(G) Wang, H. (1968) *An Outline of Human Embryology* (London: Heinemann)
(A) Weiner, N. (1948) *Cybernetics; or, Control and Communication in the Animal and the Machine* (New York: Wiley)
(A) Willier, B. H., Weiss, P. A., and Hamburger, V. (eds) (1955) *Analysis of Development* (Toronto: Saunders)
(G) Wolfe, S. L. (1972) *Biology of the Cell* (Belmont, California: Wadsworth)

Selected references to original papers

Balfour-Lynn, S. (1956) 'Parthenogenesis in human beings', *Lancet*, **1**, 1071–2.
Barrow, M. V. (1971) 'A brief history of teratology to the early 20th century', *Teratology*, **4**, 119–30.
Beck, F. (1976) 'Model systems in teratology', *British Medical Bulletin*, **32**, 53–8.
Bickers, W. (1960) 'Sperm migration and uterine contractions', *Fertility and Sterility*, **11**, 286–90.
Billingham, R. E. (1964) 'Transplantation immunity and the maternal–fetal relation', *New England Journal of Medicine*, **270**, 667–72.
Bradbury, S., Billington, W. D., and Kirby, D. R. S. (1965) 'A histochemical and electronmicroscopical study of the fibrinoid of the mouse placenta', *Journal of the Royal Microscopical Society*, **84**, 199–211.
Briggs, R., and King, T. J. (1952) 'Transplantations of living nuclei from blastula cells into enucleated frog's eggs', *Proceedings of the National Academy of Science* (Washington), **38**, 455–63.
Bulmer, D. (1957) 'The development of the human vagina', *Journal of Anatomy* (London), **91**, 490–509.
Burdi, A. R., and Faist, K. (1967) 'Morphogenesis of the palate in normal human embryos with special emphasis on the mechanisms involved', *American Journal of Anatomy*, **120**, 149–59.
Byers, B., and Porter, K. R. (1964) 'Oriented microtubules in elongating cells of the developing lens rudiment after induction', *Proceedings of the National Academy of Science* (Washington), **52**, 1091–9.
Cameron, G., and Chambers, R. (1938) 'Direct evidence of function in kidney of early human fetus', *American Journal of Physiology*, **123**, 482–5.
Carr, D. H. (1971) 'Genetic basis of abortion', *Annual Review of Genetics*, **5**, 65–80.
Carter, C. O. (1969) 'Spina bifida and anencephaly: a problem in genetic–environmental interaction', *Journal of Biosocial Science*, **1**, 71–83.
Carter, C. O. (1976) 'Genetics of common single malformations', *British Medical Bulletin*, **32**, 21–6.
Chiquoine, A. D. (1960) 'The development of the zona pellucida of the mammalian ovum', *American Journal of Anatomy*, **106**, 149–70.

Conel, J. L. (1942) 'The origin of the neural crest', *Journal of Comparative Neurology*, **76**, 191–215.
Corner, G. W. (1929) 'A well preserved human embryo of 10 somites', *Contributions to Embryology*, **20**, 80–102.
Corner, G. W. (1955) 'The observed embryology of human single-ovum twins and other multiple births', *American Journal of Obstetrics and Gynecology*, **70**, 933–51.
Coulombre, A. J. (1961) 'Cytology of the developing eye', *International Review of Cytology*, **11**, 161–94.
Crawford, J. M. (1962) 'Vascular anatomy of the human placenta', *American Journal of Obstetrics and Gynecology*, **84**, 1543–67.
De Martino, C., and Zamboni, L. (1966) 'A morphologic study of the mesonephros of the human embryo', *Journal of Ultrastructure Research*, **16**, 399–427.
Davis, C. L. (1923) 'Description of a human embryo having 20 paired somites', *Contributions to Embryology*, **15**, 1–51.
Di Virgilio, G., Lavenda, N., and Worden, J. L. (1967) 'Sequence of events in neural tube closure and the formation of neural crest in the chick embryo', *Acta Anatomica*, **68**, 127–46.
Duckett, S. (1968) 'The germinal layer of the growing human brain during early fetal life', *Anatomical Record*, **161**, 231–46.
Eakin, R. M. (1949) 'The nature of the organizer', *Science*, **109**, 195–7.
Enders, A. C. (1965) 'A comparative study of the fine structure of the trophoblast in several haemochorial placentae', *American Journal of Anatomy*, **116**, 29–67.
Fischberg, M., Gurdon, J. B., and Elsdale, T. R. (1959) 'Nuclear transfer in amphibia and the problems of the potentialities of the nuclei of differentiating tissues', *Experimental Cell Research*, Supplement **6**, 161–78.
Frederickson, R. G., and Low, F. N. (1971) 'The fine structure of perinotochordal microfibrils in control and enzyme-treated chick embryos', *American Journal of Anatomy*, **130**, 347–76.
Gerbie, A. B., Nadler, H. L., and Gerbie, M. V. (1971) 'Amniocentesis in genetic counselling. Safety and reliability in early pregnancy', *American Journal of Obstetrics and Gynecology*, **109**, 765–70.
Gierer, A. (1974) 'Hydra as a model for the development of biological form', *Scientific American*, **231** (December), 44–54.
Gillman, J. (1948) 'The development of the gonads in Man, with a consideration of the role of foetal endocrines and the histogenesis of ovarian tumours', *Contributions to Embryology*, **32**, 81–131.
Glenister, T. W., and Hamilton, W. J. (1963) 'The embryology of sexual differentiation in relation to the possible effects of administering steroid hormones during pregnancy', *Journal of Obstetrics and Gynaecology of the British Commonwealth*, **70**, 13–19.
Granholm, N. H., and Baker, J. R. (1970) 'Cytoplasmic microtubules and the mechanism of avian gastrulation', *Developmental Biology*, **23**, 563–84.
Gregg, N. H. (1941) 'Congenital cataract following German measles in the mother', *Transactions of the Ophthalmologists Society of Australia*, **3**, 35–46.
Grobstein, C. (1956), 'Inductive tissue interaction in development', *Advances in Cancer Research*, **4**, 187–236.

Hadek, R. (1965) 'The structure of the mammalian egg', *International Review of Cytology,* **18,** 29−71.
Hamburger, V., and Hamilton, H. L. (1951) 'A series of normal stages in the development of the chick embryo', *Journal of Morphology,* **88,** 49−92.
Hamilton, W. J. (1949) 'Early stages of human development', *Annals of the Royal College of Surgeons,* **4,** 281−94.
Hamilton, W. J., and Boyd, J. D. (1960) 'Development of the placenta in the first three months of gestation', *Journal of Anatomy* (London), **94,** 297−328.
Handel, M. A., and Roth, L. E. (1971) 'Cell shape and morphology of the neural tube: implications for microtubule function', *Developmental Biology,* **25,** 78−95.
Hertig, A. T., Adams, E. C., and Mulligan, W. J. (1954) 'On the preimplantation stages of the human ovum: a description of four normal and four abnormal specimens ranging from the second to the fifth day of development', *Contributions to Embryology,* **35,** 199−220.
Hertig, A. T., and Rock, J. (1945) 'Two human ova of the pre-villous stage, having a developmental age of about seven and nine days respectively', *Contributions to Embryology,* **31,** 65−84.
Heuser, C. H. (1930) 'A human embryo with 14 pairs of somites', *Contributions to Embryology,* **21,** 135−54.
Hicks, S. P., and D'Amato, C. J. (1966) 'Effects of ionizing radiations on mammalian development', *Advances in Teratology,* **1,** 195−250.
Hope, J. (1965) 'The fine structure of the developing follicle of the rhesus ovary', *Journal of Ultrastructure Research,* **12,** 592−610.
Karfunkel, P. (1971) 'The role of microtubules and microfilaments in neurulation in *Xenopus*', *Developmental Biology,* **25,** 30−56.
Knox, E. G. (1972) 'Anencephalus and dietary intakes', *British Journal of Preventive and Social Medicine,* **26,** 219−23.
Langman, J., Guerrant, R. L., and Freeman, B. G. (1966) 'Behaviour of neuroepithelial cells during closure of the neural tube', *Journal of Comparative Neurology,* **127,** 399−412.
Langman, J., and Haden, C. C. (1970) 'Formation and migration of neuroblasts in the spinal cord of the chick embryo', *Journal of Comparative Neurology,* **138,** 419−31.
Laurence, K. M., and Gregory, P. (1976) 'Prenatal diagnosis of chromosome disorders', *British Medical Bulletin,* **32,** 9−15.
Liu, H. M., and Potter, E. L. (1962) 'Development of the human pancreas', *Archives of Pathology,* **74,** 439−51.
McKay, D. G., Hertig, A. T., Adams, E. C., and Danziger, S. (1953) 'Histochemical observations on the germ cells of human embryos', *Anatomical Record,* **117,** 201−20.
Morton, W. R. M. (1949) 'Two early human embryos', *Journal of Anatomy* (London), **83,** 308−14.
O'Rahilly, R. (1966) 'The early development of the eye in staged human embryos', *Contributions to Embryology,* **38,** 1−42.
Payne, F. (1924) 'General description of a 7-somite human embryo', *Contributions to Embryology,* **16,** 117−24.
Poswillo, D. (1976) 'Mechanisms and pathogenesis of malformation', *British Medical Bulletin,* **32,** 59−64.

Renwick, J. H. (1972) 'Hypothesis: anencephaly and spina bifida are usually preventable by avoidance of a specific but unidentified substance present in certain potato tubers', *British Journal of Preventive and Social Medicine*, **26**, 67–88.

Rock, J., and Hertig, A. T. (1948) 'The human conceptus during the first two weeks of gestation', *American Journal of Obstetrics and Gynecology*, **55**, 6–17.

Ryan, K. J. (1962) 'Hormones of the placenta', *American Journal of Obstetrics and Gynecology*, **84**, 1695–713.

Saxen, L., and Toivonen, S. (1961) 'The two-gradient hypothesis in primary induction. The combined effect of two types of inductors mixed in different ratios', *Journal of Embryology and Experimental Morphology*, **9**, 514–33.

Smith, S. M., and Penrose, L. S. (1955) 'Monozygotic and dizygotic twin diagnosis', *Annals of Human Genetics*, **19**, 273–89.

Smithells, R. W. (1976) 'Environmental teratogens of man', *British Medical Bulletin*, **32**, 27–33.

Tiedemann, H. (1967) 'Inducers and inhibitors of embryonic differentiations: their chemical nature and mechanism of action', *Experimental Biology and Medicine*, **1**, 8–21.

Trelstad, R. L., Hay, E. D., and Revel, J. P. (1967) 'Cell contact during early morphogenesis in the chick embryo', *Developmental Biology*, **16**, 78–106.

Weiss, P. (1950) 'Perspectives in the field of morphogenesis', *Quarterly Review of Biology*, **25**, 177–98.

Wells L. J. (1954) 'Development of the human diaphragm and pleural sacs', *Contributions to Embryology*, **35**, 107–34.

Witschi, E. (1948) 'Migration of the germ cells of human embryos from the yolk sac to the primitive gonadal folds', *Contributions to Embryology*, **32**, 67–80.

Index

Page numbers with (g) after them refer to the glossary.

Abnormal genes, 115
Acrosome, 20, 104, 161(g)
Actinomysin, 140, 158
Adenine, 133
Adolescence, 6
Adulthood, 6
Alar plate, 39, 40, 43, 161(g)
Alkaline phosphatase, 75
Allantois, 27, 61, 70, 79, 80, 161(g)
Alpha-fetoprotein, 124
Alveoli, 71, 100
Ameloblasts, 63
Aminopterin, 116, 127
Amniocentesis, 93, 123, 161(g)
Amnion, 16, 22, 24, 27, 92–3
Amniotic cavity, 10, 27, 80, 93, 161(g)
Amniotic fluid, 71, 93, 100, 102, 117
Ampulla, 59
Anal canal, 70
Anencephalus, 93, 118, 124
Anophthalmia, 59
Anterior primary ramus, 73
Antibodies, 23, 117, 118
Antigens, 23
Aorta, 50, 51, 52, 53
Aortic arch, 50
Aortic trunk, 48, 49
Appendix, 68, 69
Arachnoid mater, 41
Aristotle, 1, 2, 111
Ascending colon, 68
Atomic bomb, 117
Atresia, 84, 106, 162(g)
Atrium, 46, 47, 48
Auditory placode, 59
Auditory vesicle, 59, 60
Automata net, 142–3
Autosome, 18, 81, 108, 162(g)
Axon, 40

Basal lamina, 29
Basal plate, 39, 40, 43, 162(g)
Base pairs, 133
Bile capillaries, 67
Biogenetic law, 3, 4

Birth, 5, 16, 76, 99–100
Bit (of information), 143
Bladder, 61, 123
Blastocyst, 5, 9, 10, 18, 22, 162(g)
Blastomere, 4, 24, 162(g)
Blood islands, 44, 162(g)
Blood vessels, 44, 50
Body stalk, 11, 27, 44, 93, 94, 95
Bone marrow, 75
Brain, 15, 37, 38, 41–4
Breech presentation, 99
Broad ligament, 85, 87
Bronchi, 70, 71
Buccopharyngeal membrane, 30, 31, 44, 45, 61, 62
Bulbus cordis, 46, 47, 48, 162(g)

Caecum, 68, 69
Calcitonin, 65
Cancellus bone, 75
Cancer, 6, 23, 155
Cardiac loop, 47
Cardinal veins, 46, 47, 51
Carotid arteries, 50
Cauda equina, 41
Cell death, 11, 117, 128, 129, 150, 151
Cell division, 11, 16, 104, 108, 139
Cementum, 63
Central canal of spinal cord, 41
Centriole, 19, 104, 105
Centrum, 77
Cerebellum, 41, 43
Cerebral aqueduct, 41, 42, 43
Cerebral cortex, 44, 123
Cerebral hemispheres, 44
Cerebrospinal fluid, 41, 43, 122, 162(g)
Cervical flexure, 43
Cervical plug, 95
Chemotaxis, 19
Child, C. M., 156
Chloramphenicol, 140
Chorioepithelioma, 23, 100
Chorion, 9, 16, 24, 94–8, 162(g)
Chorionic gonadotrophin, 27, 107
Choroid, 58

173

Choroid fissure, 57, 162(g)
Choroid plexus, 41
Chromatid, 19, 108, 109, 110, 134, 162(g)
Chromosome, 16, 18, 81, 108, 109, 116, 132, 133, 134
Ciliary body, 58
Cistrons, 133
Clavicle, 78
Cleavage, 9, 19, 144, 145, 154, 162(g)
Cleft lip, 56, 113
Cleft palate, 56, 113, 115
Clitoris, 86
Cloaca, 70, 78, 83
Cloacal membrane, 30, 31, 61, 70, 85
Cochlea, 59
Codon, 136, 139
Collecting tubules, 79
Common hepatic duct, 67
Compact bone, 75, 76
Complexity, 147
Congenital abnormalities, 111–30, 162(g)
Congenital cataract, 59
Conjoined twins, 24, 25, 162(g)
Contractions, 99
Copula, 66
Cornea, 58
Coronary sinus, 48
Corpus albicans, 107
Corpus luteum, 98, 105, 107, 162(g)
Costal processes, 77
Cotyledons, 96
Critical period, 12, 119, 162(g)
Crytorchid testes, 104
Cybernetics, 142, 162(g)
Cyclopia, 58, 59, 111, 162(g)
Cystic duct, 68
Cytogenes, 144, 162(g)
Cytosine, 133
Cytotrophoblast, 10, 94, 95, 162(g)

Decidua, 94, 95, 100, 162(g)
Deciduous teeth, 63
Dedifferentiation, 6, 8, 156
Dendrites, 40
Dental buds, 62, 63
Dental lamina, 62
Dental papilla, 62
Dental sac, 63
Dermatome, 72, 73, 162(g)
Dermis, 73
Descending colon, 70
Descent of gonads, 86–7
Development field, 33, 154–8, 163(g)
Developmental mechanics, 131
Diaphragm, 45, 100
Diaphysis, 75
Diencephalon, 41, 43
Dietary factors, 118, 125
Differentiation, 4, 9, 120, 136, 138, 143, 144, 147, 148, 163(g)
Digestive system, 11, 12, 61–70, 101–2
Diploid, 108, 163(g)
Distal convoluted tubule, 80
Division of labour, 17
Dizygotic twins, 24, 26, 163(g)
DNA, 133–7, 163(g)
Dorsal aortae, 46, 51

Dorsal lip of blastopore, 32
Dorsal pancreatic bud, 68
Dorsal root, 40
Dorsal root ganglia, 40, 163(g)
Dose rate, 120
Down's syndrome, 113, 114
Driesch, 4
Drugs, 116
Ductus arteriosus, 50, 53, 101, 163(g)
Ductus venosus, 51, 101, 163(g)
Duodenum, 67
Dura mater, 41

Ear, 14, 15, 59–60
Ectoderm, 10, 11, 27, 29
Ectopic pregnancy, 23, 163(g)
Efferent ductules, 83, 104
Einstein, A., 151
Ejaculatory duct, 104
Electron microscopy, 19, 33, 163(g)
Embryonic disc, 10, 27, 163(g)
Embryonic period, 5, 10
Enamel, 63
End bud, 38
Endocardial cushions, 47, 48, 163(g)
Endocardial tube, 45, 163(g)
Endoderm, 10, 11, 12, 27, 29, 61, 70
Endometrium, 22, 94, 107, 163(g)
Environment, 17, 113, 115, 118, 119, 124, 156
Ependyma, 39, 40, 41
Epidemiology, 112, 163(g)
Epidermis, 73
Epididymis, 83, 104, 105
Epigenesis, 2, 143, 144, 163(g)
Epimere, 73
Epiphysial plate, 75
Epiphysis, 75
Epispadias, 86
Experimental embryology, 4, 5, 57, 88, 119–20, 131, 140, 144, 145
External ear, 59, 64
External genitalia, 15, 85–6
Extraembryonic coelom, 10, 27, 34, 162(g; under coelom)
Extraembryonic membranes, 92–8, 120
Extraembryonic mesoderm, 27, 94
Eye, 13, 15, 56–9

Fabricus, 2
Face, 53–6
Fertilization, 8, 18, 19, 110, 148, 154, 163(g)
Fimbria, 84, 110
Foetal circulation, 51–3
Foetal haemolysis, 117
Foetal period, 6, 15, 88–92
Foeto/maternal relationship, 22, 23
Folding of cell layers, 11, 29, 30, 35, 36, 57, 59, 84
Folic acid, 127
Follicle, 26, 84, 105, 106
Follicle-stimulating hormone, 26, 106, 107
Follicular antrum, 106
Fontanelle, 76, 163(g)
Foramen ovale, 48, 49, 101, 163(g)
Forebrain, 41
Foregut, 61, 66, 67
Frontonasal process, 53

Gallbladder, 68
Gametes, 8, 18, 103–10, 148, 149, 150, 163(g)
Gametogenesis, 83, 103–10
Genes, 133, 136, 140–1, 164(g)
Genetic code, 136–7
Genetic sex, 18, 81
Genital folds, 85
Genital ridge, 82
Genital swellings, 85, 86
Genital system, 81–8
Genital tubercle, 85
German measles, 115
Germ layers, 3, 11, 27, 29, 164(g)
Gierer, A., 156
Glial cells, 39
Glomerular capsule, 80
Glomerulus, 80
Goette, 4
Gonadotrophic hormone, 83, 98
Gonads, 12, 81, 86, 117
Graafian follicle, 106, 164(g)
Gravitation, 152
Greater curvature, 67
Greater omentum, 67
Greeks, 1
Grey matter, 39, 43
Growth, 15, 88–92
Guanine, 133
Gubernaculum, 86, 87, 164(g)

Haeckel, 3, 4
Haemopoesis, 67, 75
Haploid, 108, 164(g)
Harrison, 5, 156
Harvey, William, 2
Heart, 11, 44–7, 66
Hensen's node, 29, 30, 32
Hepatic bud, 67
Hill analogy, 148–51
Hindbrain, 41, 59, 122, 129
Hindgut, 61, 70
His, 4
Histones, 133, 134, 164(g)
Hole in the heart, 49
Homologous pairs, 108, 136
Hormones, 16, 18, 26, 88, 97, 98, 99, 106, 107, 116
Horseshoe kidney, 80, 81
Hydramnios, 93, 164(g)
Hydrocele, 87
Hydrocephalus, 42, 122–3, 129, 164(g)
Hyoid bone, 64
Hypomere, 73
Hypospadias, 86

Ileum, 68, 69
Immune response mechanism, 22, 117
Imperforate anus, 70
Implantation, 5, 9, 22, 164(g)
Indifferent period, 12, 82, 164(g)
Induction, 5, 32–3, 56, 59, 80, 131, 140, 150, 151, 158, 164(g)
Inevitability, 157
Infection, 115
Inferior mesenteric artery, 81
Inferior vena cava, 52
Influenza, 115, 125, 127

Information, 16, 133, 136, 143–5, 154, 164(g)
Inguinal canal, 87
Inner cell mass, 9, 10, 22, 24, 27, 164(g)
Insulin, 68, 116
Intermediate mesoderm, 34, 72, 78, 79, 164(g)
Internal ear, 59
Interventricular septum, 48
Intervertebral disc, 77
Intervillous space, 96, 97
Intraembryonic coelom, 34, 45, 70, 82, 162(g; under coelom)
Iris, 58
Irradiation, 116, 127

Jacob and Monod, 140, 142
Jejunum, 68, 69
Joints, 78

Kidneys, 78, 80, 81, 102
Klinefelter's syndrome, 114

Labia majora, 87
Labia minora, 86
Labour, 99, 164(g)
Lacunae, 94
Lanugo, 89, 91
Larynx, 64, 70
Lateral lingual swellings, 66
Lens fibres, 58
Lens placode, 56, 57
Lens vesicle, 56, 58
Lesser curvature, 67
Lesser omentum, 67
Leukaemia, 8, 156
Ligamentum arteriosum, 101
Ligamentum venosum, 101
Limb buds, 13, 74, 78
Limb girdles, 78
Liver, 62, 67, 69
Loop of Henle, 80
Lung buds, 70

Malpighi, 2
Mandible, 64
Mandibular process, 53, 64
Mangold, 5
Mantle layer, 39, 164(g)
Manx cat, 127
Marginal layer, 39, 40, 164(g)
Maternal impressions, 1
Maturation, 15, 88
Maxillary process, 53, 56, 64
Mechanical pressure, 117
Meconium, 102
Medulla oblongata, 41
Meiosis, 104, 106, 108, 165(g)
Membranous labyrinth, 59, 165(g)
Meninges, 41, 129, 165(g)
Menstrual flow, 108
Mesenteries, 67, 68
Mesoderm, 10, 11, 28, 29, 44, 64, 71, 72
Mesonephric duct, 78, 79, 83
Mesonephric tubules, 78, 79, 83
Mesonephros, 78, 79, 165(g)
Messenger RNA, 137, 139, 140, 158
Metanephrogenic caps, 79, 80
Metanephros, 78, 79, 165(g)

Microfilaments, 35, 36, 57, 127, 128, 165(g)
Microphthalmia, 59
Microtubules, 19, 28, 32, 35, 36, 56, 105, 127, 128, 165(g)
Midbrain vesicle, 41
Middle ear, 59, 64
Midgut, 61, 67, 68—70
Minkowski, 151
Mitochondria, 144—5
Mitosis, 108, 109, 165(g)
Mitotic spindle, 19, 108, 165(g)
Model building, 141—2, 165(g)
Mongolism, 113, 114
Monozygotic twins, 24, 125, 165(g)
Morton, W. R. M., 122
Mouth, 53, 61, 62—3
Mullerian ducts, 84
Multipolar neurons, 40
Muscles, 12, 15, 73—4
Musculoskeletal system, 72—8
Myelination, 39, 165(g)
Myotome, 72, 73, 165(g)

Nasal cavity, 54
Nasal septum, 54
Neonatal period, 100—3, 165(g)
Nephron, 80
Nerve plexuses, 74
Nervous system, 35—44, 103
Neural arch, 77
Neural crest, 35, 36, 40, 128, 165(g)
Neural folds, 12, 37, 128
Neural plate, 11, 12, 30, 32, 33, 56, 128, 165(g)
Neural retina, 56, 58
Neural tube, 13, 35, 38, 39, 126
Neurocytes, 39, 40
Neurulation, 30, 35—8
Newborn baby, 103
Newton, Isaac, 151
Noise, 143, 165(g)
Normal distribution, 112
Notochord, 30, 31, 32, 33, 35, 76, 165(g)
Nucleolus, 137
Nucleus pulposus, 30, 77

Oblique facial cleft, 56
Odontoblasts, 62
Oesophagus, 66
Oestrogen, 98, 106, 115
Olfactory pits, 53, 54
Oligohydramnios, 93, 165(g)
Oocytes, 105, 106, 108
Oogenesis, 105—7, 114, 165(g)
Oogonia, 84, 105, 108
Operator genes, 140—1
Operon concept, 140—1, 165(g)
Optic cup, 56, 57, 165(g)
Optic nerve, 56
Optic vesicle, 56, 57, 165(g)
Oral contraceptive, 107
Organizer, 32, 131, 155, 165(g)
Organ of Corti, 59
Ossicles, 60, 64, 165(g)
Ossification, 12, 64, 74—6, 165(g)
Ossification centres, 75, 78
Osteoblasts, 75
Osteoclasts, 75

Osteocytes, 75
Osteoid, 75
Ovary, 26, 84, 105
Overgrowth, 128
Ovulation, 106, 110
Ovum, 18, 19, 20, 82, 105, 106, 109, 148, 149
Oxytocin, 99

Palatal processes, 53
Pancreas, 62, 68
Pancreatic ducts, 68
Pancreatic islets, 68
Pander, 3
Paramesonephric duct, 83, 84, 85
Parasitic twin, 24
Parathyroid glands, 62, 65
Parthenogenesis, 148
Penis, 86
Pericardial cavity, 45, 47
Periodontal ligament, 63
Periods of development, 5
Periosteum, 75
Permanent teeth, 63
Phallus, 85
Pharyngeal arches, 13, 50, 63—6, 73, 166(g)
Pharyngeal clefts, 63, 64
Pharyngeal pouches, 63, 64—5
Pharyngotympanic tube, 62, 64
Pharynx, 63—6, 166(g)
Physiological gradient, 33, 156
Physiological herniation, 68, 164(g; under herniation)
Pia meter, 41
Pigment layer of retina, 58
Pineal body, 43
Pinna, 60, 64
Pituitary gland, 43
Placenta, 9, 11, 22, 24, 26, 51, 53, 80, 95—8, 100, 166(g)
Placental infarcts, 98
Placental insufficiency, 99
Placental septae, 96
Placenta praevia, 98
Plasma, 44
Pleural cavities, 70
Polar body, 106, 109, 166(g)
Polio virus, 136
Polycystic kidney, 81
Pons, 41
Posterior primary ramus, 73
Postnatal period, 5
Potato blight, 125
Potential, 147—50, 166(g)
Preformation, 2, 143, 166(g)
Pre-implantation period, 5, 8
Prematurity, 26, 92
Prenatal detection, 123—4
Prenatal period, 5, 166(g)
Prepubertal period, 6
Prevention, 124
Primary oocytes, 84
Primitive streak, 11, 24, 28, 33, 166(g)
Primordial germ cells, 82, 84, 166(g)
Probe patency, 101
Processus vaginalis, 87
Progesterone, 97, 99, 107, 116
Pronephros, 78, 79, 166(g)

Prostate gland, 62, 83
Protein synthesis, 139
Proximal convoluted tubule, 80
Pseudostratified epithelium, 38
Pulmonary arteries, 50
Pulmonary trunk, 48, 49
Pulmonary veins, 101
Pulp, 63
Pyloric stenosis, 67

Quadruplets, 26

Rathke, 3
Rauber, 4
Rectum, 70
Reduction of chromosomes, 18, 108
Redundancy, 136, 143, 166(g)
Reflex activity, 15, 16, 54, 71, 80, 93
Regulator genes, 140–1
Renwick, J. H., 125
Replication, 19, 133–6, 166(g)
Repressor, 140–1
Reproductive system, 12
Respiratory system, 11, 62, 70–2, 100
Rete testis, 83, 104
Rhesus sensitization, 118
Ribosomal RNA, 137, 139
Ribosomes, 137, 138, 139
RNA, 136, 137–41, 166(g)
Root sheath, 63
Rotation of midgut, 69
Round ligament of ovary, 87
Round ligament of uterus, 87
Roux, 4
Rubella, 115

Saccule, 59
Sclera, 58
Sclerotome, 72, 73, 76, 77, 166(g)
Scrotum, 86, 87, 104
Semicircular canals, 59, 60
Seminal vesicle, 83, 104
Seminiferous tubules, 83, 87, 104, 105
Septation of heart, 47, 48
Septum primum, 47, 48, 166(g)
Septum secundum, 47, 48, 166(g)
Septum transversum, 45, 47, 51, 66, 69
Sertoli cells, 83, 104
Sex chromosomes, 18, 81, 108, 114, 166(g)
Siamese twins, 24
Sigmoid colon, 70
Sinus venosus, 46, 47, 48, 51, 52, 166(g)
Skeletal system, 12, 74–8
Skull, 76, 100
Somatic mesoderm, 34, 72, 166(g)
Somites, 12, 15, 33, 72–3, 129, 166(g)
Spacetime, 151
Specialized intercellular junctions, 29, 30, 33, 38
Spemann, H., 5, 32, 156
Spermatids, 104
Spermatocytes, 104, 108
Spermatogenesis, 103–5, 166(g)
Spermatogonia, 83, 104, 108
Spermatoza, 18, 19, 20, 104–5, 110
Spina bifida, 121–30
Spinal cord, 39, 40, 41, 121
Spinal nerves, 40, 73, 74

Spiral ganglion, 59
Spiral septum, 48, 49, 166(g)
Spontaneous abortion, 113–4
Stockard, 119
Stomach, 67, 69
Structural genes, 140–1
Subclavian artery, 50
Superior mesenteric artery, 68, 69
Superior vena cava
Suprarenal glands, 81
Surfactant, 72, 100, 166(g)
Suspensory ligament, 58
Syncytiotrophoblast, 10, 94, 95, 166(g)
Syphilis, 116

Tail flexure, 43
Teeth, 62
Telencephalon, 41
Teratogen, 12, 119, 120, 154, 155, 167(g)
Teratology, 111, 119–20, 127, 131, 167(g)
Testis, 81–3, 86, 103–4
Testosterone, 87, 88, 104
Tetralogy of Fallot, 49, 167(g)
Thalidomide, 116
Thymine, 133
Thymus gland, 62, 65
Thyroid gland, 62, 65–6
Tongue, 54, 65, 66, 73
Tonsil, 65
Toxoplasma gondii, 115
Trachea, 70
Transfer RNA, 137, 139
Transport of gametes, 110
Transverse colon, 68, 70
Triplets, 26
Trisomy, 114
Trophoblast, 10, 22, 23, 27, 44, 94, 167(g)
Trophoblastic villi, 94–5
Trypan blue, 127
Tuberculum impar, 66
Tunica vaginalis, 87
Turner's syndrome, 114–5
Tympanic membrane, 60, 64

Ultrasound scanning, 124
Umbilical cord, 11, 16, 93, 97, 101, 103, 167(g)
Umbilical vessels, 11, 46, 51, 53, 67, 81
Uracil, 137
Ureter, 79, 81
Ureteric bud, 79, 80
Urethra, 83, 85, 104
Urinary system, 12, 78–81, 102
Urogenital sinus, 70, 83, 84, 85
Urorectal septum, 70, 167(g)
Uterine tube, 9, 19, 84, 85, 110
Uterus, 9, 22, 84, 85, 99, 107, 110
Utricle, 59

Vagina, 84, 85, 99, 110
Vas deferens, 83, 104
Ventral pancreatic bud, 68
Ventral roots, 40
Ventricles of brain, 41, 43
Ventricles of heart, 46, 47, 48
Vernix caseosa, 89, 91
Vertebrae, 76, 77
Vertebral column, 76, 77, 121

Vestibular apparatus, 59
Villous tree, 96
Visceral mesoderm, 34, 44, 72, 167(g)
Vital centres, 44
Vitamin A, 120, 127
Vitelline arteries, 46
Vitelline veins, 46, 51, 67
Von Baer, 3

Waddington, C. H., 148
Waterbed analogy, 152
Waters, 93

Watson and Crick, 133, 142
Weiner, 142
White matter, 39, 43
Wolff, 2
Wolffian duct, 78, 79, 83

X-rays, 116, 127, 133

Yolk sac, 10, 22, 27, 44, 61, 70, 167(g)

Zona pellucida, 9, 18, 19, 22, 106, 167(g)
Zygote, 5, 16, 17, 19, 26, 144, 149, 167(g)